Modern
Scientific Thought

P.C. CHANDRASEKHARAN

INDIA • SINGAPORE • MALAYSIA

Notion Press

Old No. 38, New No. 6
McNichols Road, Chetpet
Chennai - 600 031

First Published by Notion Press 2018
Copyright © P.C. Chandrasekharan 2018
All Rights Reserved.

ISBN 978-1-64429-144-3

This book has been published with all efforts taken to make the material error-free after the consent of the author. However, the author and the publisher do not assume and hereby disclaim any liability to any party for any loss, damage, or disruption caused by errors or omissions, whether such errors or omissions result from negligence, accident, or any other cause.

No part of this book may be used, reproduced in any manner whatsoever without written permission from the author, except in the case of brief quotations embodied in critical articles and reviews.

"To the loving memory of my wife Jaya."

Thank you for sixty years of companionship.

Contents

Preface .... *xi*
Acknowledgements .... *xiii*

Chapter 1: A Compendium of Essays on Modern Scientific Thought 1
1.1 Introduction 1

Chapter 2: Public Understanding of Science 5
2.1 Introduction 5
2.2 What is Science Literacy? 5
2.3 The First Step 5
2.4 The Case of Skilled and Semi-skilled Labour 6
2.5 Educating the Bureaucrats, Politicians and Legislators 6
2.6 The Voice of the People 6
2.7 Role of Science Education in Schools 7

Chapter 3: Origin & Evolution of the Universe – The Grand Design 9
3.1 Historical Perspective 9
3.2 The Evolutionary Process – Step By Step Description 10
3.3 The Cosmic Microwave Background Radiation (CMBR) 11
3.4 Formation of Stars and Galaxies 11
3.5 Dark Matter and Dark Energy 12
3.6 Critical Density 13
3.7 Some Frequently Asked Questions 13
3.8 A Peep into the Future 15

Chapter 4: The Microscopic World of Atoms 19
4.1 Introduction 19
4.2 Anomalies in the Rutherford Model 20
4.3 Explanation of Some Commonly Used Terminology 21
4.4 Discovery of Sub-Atomic Particles 23
4.5 The Higgs Boson 27
4.6 Summary 28

Chapter 5: The Birth and Death of Stars 29
5.1 The Cosmic Architecture 29
5.2 Birth and Death of Stars 30

5.3 Low Mass Stars 31
5.4 White Dwarfs and Neutron Stars 31
5.5 The Mystery of the Supernova 32
5.6 More about Neutron Stars, and Black Holes. 32
5.7 Quasi Stellar Radio Sources (QUASAR) 34
5.8 Conclusion 34

Chapter 6: Unification of the Forces of Nature – In search of the Holy Grail 35
6.1 Introduction. 35
6.2 Structure of the Atom – A Recapitulation 35
6.3 Gravitational Force 38
6.4 Electromagnetic Force 38
6.5 The Strong Force 39
6.6 The Weak Force 40
6.7 Unification of Forces 42

Chapter 7: Our Planet Earth and its Neighbours 45
7.1 Introduction. 45
7.2 What is a Planet? 46
7.3 The Sun and Its Immediate Neighborhood 46
7.4 Dwarf Planets, Asteroids and Comets. 52
7.5 Exoplanets 54

Chapter 8: Scanning the Universe 57
8.1 An Overview 57
8.2 Observational Astronomy – Some Important Milestones 59
8.3 The Large Hadron Collider 62
8.4 Gravitational Waves and Neutrino Astronomy. 63
8.5 Summary 64

Chapter 9: Nuclear Fission Versus Nuclear Fusion 67
9.1 Introduction. 67
9.2 Some Basic Preliminaries 68
9.3 Nuclear Fission Reactor 71
9.4 Nuclear Fusion Reactor 73
9.5 Fission Versus Fusion – A Comparison 76
9.6 Conclusion 77

Chapter 10: Relativistic Theory 79
10.1 Introduction. 79
10.2 Newtonian Mechanics. 79
10.3 Special Theory of Relativity 80
10.4 Time Dilation and Length Contraction 81

10.5 Mass Energy Equivalence 83
10.6 General Relativity 84

Chapter 11: Quantum Mechanics – A Paradigm Shift 87
11.1 Introduction 87
11.2 Classical Physics Versus Quantum Mechanics 87
11.3 De Broglie Waves 89
11.4 The Bohr Model of the Atom 89
11.5 Heisenberg's Uncertainty Principle 91
11.6 The Wave Function and Schrödinger's Equation 92
11.7 Feynman's Path Integral Formulation 94

Chapter 12: Biology's Big Bang 97
12.1 Early Developments 97
12.2 Introduction to Molecular Biology 97
12.3 Replication of the DNA 99
12.4 Structure of Chromosomes 100
12.5 Messenger RNA (mRNA) and Ribosome 102
12.6 The Idea of Epigenetics 103

Chapter 13: Synthetic Biology – Playing God! 105
13.1 Introduction 105
13.2 Recapitulation of Some Basic Concepts 105
13.3 Replication, Regulation and Recombination of Genes 107
13.4 Stem Cells, Embryonic Cells and Induced Pluripotent Stem cells (iPSC) 108
13.5 Cloning and Gene Therapy 110
13.6 Gene Therapy 110
13.7 Genetically Linked Diseases 112
13.8 DNA Editing Techniques – CRISPR- Cas9 and Beyond 113

Chapter 14: Global Warming and Climate Change 115
14.1 Introduction 115
14.2 What is Global Warming? 115
14.3 Deleterious Consequences of Green House Gas Emission 116
14.4 The Ozone Depletion Problem 117
14.5 Strategies to Control Global Warming 117
14.6 Some Implementable Damage Control Measures 119
14.7 Conclusion 120

Chapter 15: Artificial Intelligence – Cutting Edge of Modern Technology 121
15.1 Artificial Intelligence – An Overview 121
15.2 Classification of Artificial Intelligence 121
15.3 Test for Computer Intelligence 122

15.4 Neuron Cells Architecture 122
15.5 Creating Artificial Neural Networks (ANN) 124
15.6 Multilevel Neuron Networks 125
15.7 Major Artificial Intelligence Areas 126
15.8 Evolutionary Computation and Genetic Algorithms. 127
15.9 Future of Artificial Intelligence 128

Chapter 16: Explosive Growth of Information Technology and
                Communication Systems 131
16.1 The Rise and Rise of Information Technology. 131
16.2 Historical Perspective 132
16.3 Optical Fiber Communications 133
16.4 The Electronics Revolution 134
16.5 The Internet and the World Wide Web 134
16.6 Web Search Engines 135
16.7 Internet: Social Media 135
16.8 Packet Switching 136
16.9 Conclusion 136

Chapter 17: Number Theory – Unfolding a Fascinating Tapestry 139
17.1 Introduction 139
17.2 Recapitulation of Basic Terminology 139
17.3 Distribution of Prime Numbers 140
17.4 The Prime Number Theorem 142
17.5 The Twin Prime Gap Conjecture 143
17.6 The Mersenne Number 143
17.7 Fermat's Last Theorem 144
17.8 Applied Number Theory 144

Chapter 18: Limits of Science 147
18.1 The Methodology of Science 147
18.2 An Alternative View Point 147
18.3 The Anthropic Principle 148
18.4 Science Versus Religion 148
18.5 Believers and Non-Believers 149
18.6 Conclusion 150

Glossary *151*
Index *169*

LIST OF ILLUSTRATIONS

Chapter 3
Fig 1: Break up of the constituents of the universe 16
Fig 2: A graphical representation of the expansion of the universe 16

Chapter 4
Fig 1: Energy Levels of the Hydrogen Atom. 23
Fig 2: Standard Model of particle physics 25
Fig 3: The inside structure of the hydrogen atom. 27

Chapter 6
Fig 1: Unification of Forces 43

Chapter 8
Fig 1: Electromagnetic Radiation over the entire frequency Spectrum 57
Fig 2: Filtering of EM waves by earth's atmosphere. 58

Chapter 9
Fig 1: Atomic number versus binding energy per nucleon 70
Fig 2: Illustration of Fusion and Fission reactions 71
Fig 3: Chain Reaction 72
Fig 4: Structure of Tokamak. 74

Chapter 10
Fig 1: Stationary and Moving clocks 81

Chapter 11
Fig 1: Feynman's Path Integral. 94

Chapter 12
Fig 1: Transition from nucleotides to proteins. 99
Fig 2: Chromosome structure in a human cell. 101

Chapter 15
Fig 1: Neuron cell and its connections. 123
Fig 2: Artificial Neural Network 124
Fig 3: A feed forward multilayer neural network. 126

Chapter 17
Fig 1: Representation of real numbers on the number line 140

LIST OF TABLES

Chapter 3
Table 1: Evolution of Universe from Big Bang to the Present Time 17

Chapter 4
Table 1: Masses of particles in different units 22

Chapter 6
Table 1: Fundamental Forces and their Interaction 37
Table 2: Leptons and Quarks. 37
Table 3: Mesons and Baryons. 37

Chapter 7
Table 1: Sun versus First Stars 47
Table 2: Essential Parameters of Planets in the Solar System. 51

Chapter 17
Table 1: Distribution of Prime Numbers for Integers ranging from 1 to 1000 141
Table 2: Comparison between predicted and actual proportion of primes. 142

Preface

What exactly constitute modern scientific thinking? It is a difficult question to answer as the response depends upon the predilections and interests of the person to whom the question is addressed.

However, there can be no disagreement regarding the inclusion of basic scientific facts in the book. Never in human history has science and its applications developed so fast as in the 20th century. This period has seen the birth of three fundamental science concepts – Relativity, Quantum Mechanics and Molecular Biology.

Taking up relativity, it is entirely the brain child of one individual – Albert Einstein. His leap of imagination has resulted in the unification of space and time, not as two separate entities but as part of the space – time continuum. Gravitational force which was earlier described by Newton as a force between masses (an idea which remained undisputed for over three centuries) was reinterpreted by Einstein as a consequence of the warping of space - time. Equally revolutionary in character was the birth of Quantum Mechanics as a result of the contributions made by a succession of brilliant scientists – Bohr, Heisenberg, Dirac and Schrodingar. Quantum Mechanics could truly be described as a triumph of mind over matter.

Another astounding development, which occurred in the mid – fifties was the birth of Molecular Biology. In earlier times, thanks to the outstanding contributions of Darwin and Mendel, biology was treated as an observational discipline. Its molecular dimension was unfolded by the historic discovery of the structure of the DNA by biologists Crick and Watson. More important developments followed, for example in the delineation of the human genome and the introduction of synthetic biology.

Truly, we are lucky that we are living in the golden age of scientific development. To sum up General Relativity helped us to understand the deep mysteries of the universe. Quantum Mechanics led to the discovery of transistors and the consequent developments ranging from computers to hand held electronic devices. Molecular biology has proved to be of immense help in fighting genetically transmitted diseases and in raising genetically modified crops.

It is against this background that I have chosen to write the book, "Modern Scientific Thought". The scientific concepts described above along with modern science applications which have a direct bearing on our everyday lives form the core of this book. The intention of the author in writing this book is to convey profound scientific ideas in a simple and easily digestible form. If this aim is even partially fulfilled, then the author would consider it as the greatest reward.

Acknowledgements

To begin with, I gratefully acknowledge the help and support, I received from various quarters in the preparation of the book. First and foremost, I thank my family members, who suffered years of benign neglect, during the period I was totally engrossed in writing the book. I want to especially thank my son Arun, for painstakingly going through the contents of the book and for his meticulous typing of the manuscript. I am deeply indebted to Prof. S.K. Patnaik of the Electrical Engineering department of Anna University for his help and encouragement. Finally, it is a pleasure to thank The Notion Press, publishers of the book, especially their energetic publishing manager Ms. Manisha M who spared no pains to meticulously edit the book and make it highly presentable.

Chapter 1

A Compendium of Essays on Modern Scientific Thought

1.1 INTRODUCTION

We are living in an era marked by explosive developments in Science & Technology. During the past hundred years, we have sent a man to the moon, unraveled the structure of the human gene and constructed the world wide web (www), to name but a few of sciences' outstanding achievements. It is no longer sufficient for the lay public to be merely conversant with reading and writing. They are now constantly exposed to scientific developments in their everyday life and therefore there is a dire need to cultivate what is commonly known as "Science literacy". Way back in 1959, the well known science philosopher and thinker, C.P Snow, in his historic essay titled "The two cultures" bemoaned the widening gulf of comprehension separating literary intellectuals from the scientific community. Indeed this gulf has now further widened, with the new divide existing even among different scientific disciplines. It is common knowledge that scientists working in one specialized area are unaware of what is happening in related scientific fields. The trend now appears to be to learn more and more of less and less.

It is this unfortunate situation that has primarily motivated the author to string together a range of science related topics under one umbrella and to further explain these topics in a simple and coherent manner. The aim is to evoke interest among scientists in areas other than their own and in this process widen their horizons. The book further attempts to target the not so well informed public, many of whom may have had only a minimal exposure to science, possibly at the high school level. To cater to such a wide audience the author has attempted to keep the discussion simple by eliminating mathematical treatment as far as possible. Admittedly, some of the topics listed require a profound appreciation of science and mathematics. Those who cannot follow such topics can rest assured that by skipping them much will not be lost. Certain nomenclature and definitions frequently referred to in the main text are further explained in the Glossary at the end of the book. The choice of the topics selected is by no means comprehensive and is based on the author's predilections and interests. To sum up, an attempt has been made to highlight some of the significant aspects of modern scientific thought, with the intent of broadening the vision and understanding of the reader. Brief descriptions of the contents of the various chapters follow.

Chapter-II deals with the public understanding of science. The level of scientific understanding expected from various sections of society varies widely. To begin with, a majority of people especially those living in rural areas have hardly any exposure or understanding of science. The aim is to educate them, for example, about the benefits accrued by following hygienic principles, which will in turn enhance their quality of life. Further, they are made aware of the fact that modern developments in medicine, such as vaccines, have almost obliterated diseases like small pox and polio from the face of the earth. This is in stark contradiction to long held superstitious beliefs that such diseases could only be cured by propitiating evil spirits. Our bureaucrats, politicians and legislators require a different type of training. They have to often deal with various types of developmental projects which could possibly have direct socio-economic and environmental impact on peoples lives. In such a situation informed decision making is of great relevance and science could certainly help in this regard. Science education among our youth is another area which requires urgent attention. Youth should be inspired by the idea that science is a great adventure and that the effervescence and excitement that scientific study generates in their minds should far outweigh all other considerations.

Chapter-III is cast in a totally different mould. Here, we are enquiring about the origin and evolution of the vast universe of which we are only an infinitesimal and insignificant part. However, standing on this speck of dust called the earth we have been able to unravel some of the deepest mysteries of the universe. This is certainly an achievement of unparalleled magnitude.

In Chapter-IV, the discussion is centered on the microscopic world of protons, neutrons, electrons and a host of other subatomic particles. What are the laws which govern the interaction between these various particles? Can they be applied without modification to the behavior of matter on a macroscopic scale? Why is it that small deviations from natural laws are totally inimical to human existence? These are deep questions to which partial answers are sought in this chapter.

In Chapter-V, we deal with yet another aspect of the universe, namely the birth and death of stars. Our universe contains billions of galaxies and each galaxy contains over a billion stars. Cosmologists have discovered that stars are not permanent fixtures in the sky, but go through the process of birth and death just as we human beings do. Only, the processes involved and the time scales are different. Such matters are studied in detail in this chapter.

Chapter-VI deals with the unification of the forces of nature. In spite of the multiplicity of forces we experience in everyday life, essentially there are only four distinct forces operating in nature. Two of these namely electromagnetic and gravitational forces are quite familiar to us. The remaining two forces in the form of strong and weak interactions are confined to inter atomic distances and therefore we hardly experience them. Nevertheless they are extremely significant and play an important role in the birth and evolution of the universe.

In Chapter-VII, the focus is on the structure and evolution of the solar system inhabited by planets, asteroids and comets. It turns out that among the planets of the solar system, earth alone supports life. What inhibits the support of life in other planets and how do these planets differ from planet earth, form the subject matter of this chapter.

Chapter-VIII, deals with a wide range of astronomical measurements and measurement methods involved. In any scientific endeavor, measurements play an important role, more so while dealing with astronomical phenomena, which challenge our ingenuity and innovative capacity.

In Chapter-IX, the subject matter dealt with has both theoretical and practical significance. The type of nuclear reaction which takes place deep inside the stars namely fusion reaction is not only of theoretical importance but also of extreme interest to us on earth, in our quest for generating a renewable and sustainable form of clean energy. However, we are yet to replicate fusion reaction here on earth on a commercially viable scale. The second type of nuclear reaction namely fission reaction has been successfully mastered by our scientists and there are quite a number of nuclear power stations on earth working on this principle. However, after the tragic accidents involving nuclear reactors in Chernobyl and Fukashima, nuclear power generation has now come under a cloud and there is a noticeable policy shift among nations to go in for other forms of renewable energy like; wind and solar power. Some of these problems are discussed in this chapter.

Twentieth century has witnessed the emergence of two path breaking scientific theories namely Relativity and Quantum Mechanics. Relativity theory is solely based on the seminal work of Einstein while a number of other brilliant scientists like Bohr, Heisenberg, Schrödinger and Dirac to name a few, have contributed to the development of Quantum theory. Elementary concepts pertaining to these two theories are described in Chapters- X and XI.

If the twentieth century belongs to the physical sciences, according to many, the twenty-first century will be dominated by biological sciences. Glimpses of the power and versatility of molecular biology have already been demonstrated in the seminal work of Francis Crick and James Watson, when they unveiled the double helical structure of the DNA in 1953. This has opened up a host of research opportunities culminating in the successful completion of the Human Genome Project in 2003. It turns out that new discoveries in synthetic biology will enable scientists to manipulate our genes, which opens up bright prospects for curing certain diseases, in the preparation of new drugs and in producing genetically modified food. These topics are discussed at some length in Chapters- XII and XIII.

Nations the world over, have now woken up to global warming and its consequent effects on climate change. Indiscriminate burning of fossil fuels and deforestation has resulted in the unbridled accumulation of green house gases in the atmosphere. To avert global warming concerted efforts by all countries are needed and any slackness in this matter is bound to lead to a major catastrophe. The physics of global warming as well as the consequences flowing out of it are dealt in Chapter-XIV.

Chapter-XV is devoted to an explanation of the so called discipline of *Artificial Intelligence*. This has permeated every aspect of human activity. It will make machines smart and efficient, so goes the theory and soon planet earth will run more efficiently without the help from humans. Different aspects of AI are discussed in this chapter.

The past hundred years have seen the explosive proliferation of information as well as the means of delivering the same on a global scale. Thanks to the discovery of radio, telecommunications, television and smart phones aided by the vast reach of the internet, instant communication is now possible. Such facilities have proved to be extremely useful in disseminating scientific information in the quickest possible time. These facts form the subject matter of Chapter-XVI.

The inclusion of a mathematics oriented topic like *Number Theory* in Chapter-XVII calls for some explanation. According to Galileo, considered to be the father of modern scientific approach, *"The book of nature is written in the language of mathematics"*. Mathematics widely acknowledged as the queen of all sciences, is characterized by clarity, rigor and precision. As a branch of mathematics, Number theory possesses all these attributes in abundant measure. Great mathematicians in the past like Gauss, Euler and Ramanujan have contributed significantly to this area. Further, numbers are familiar to all of us as we constantly encounter them in our daily transactions. This explains the rationale behind choosing Number Theory as a worthy candidate for Chapter-XVII.

Finally, in Chapter-XVIII, we discuss the limits of science. To begin with, the rationale and significance of scientific approach as applicable to a wide variety of problems is explained. A scientific theory is based on the verification of results flowing out of experiments suggested by the theory. In sharp contrast, the so called pseudo sciences have no scientific moorings and are not subject to verification. In fact their validity is based mainly on faith and intuition. This does not imply that science has answers to all the questions confronting humanity. According to the renowned British astrophysicist Arthur Eddington, theories accepted by science could only address those parts of the universe which are quantifiable. Human experiences like love, compassion and aesthetics are beyond the reach of science. Searching and not finding, forms the basis of both science and religion.

Chapter 2

PUBLIC UNDERSTANDING OF SCIENCE

"The ability to perceive or think differently is more important than knowledge gained"

– David Bohm

2.1 INTRODUCTION

We are living through momentous times. Great discoveries in science and technology have affected the lives of common folk in no uncertain manner. Science and its applications touch every facet of their existence. Use of radio, computers, cell phones, television and internet have become common place even among the rural folk living in remote areas far from the cities. There was a time when literacy meant merely ones capacity to read and write. This is no longer true. In addition to possessing these faculties, ordinary citizens, no matter to what rung of society he or she may belong have to be conversant with what is commonly known as "Science literacy".

2.2 WHAT IS SCIENCE LITERACY?

By science literacy one means the acquisition of a working knowledge in science which is absolutely required if one is to lead a meaningful life in the modern world. This by no means implies that one should be conversant with advanced knowledge in physics, chemistry or biology. What is expected is a rudimentary knowledge of general science to intelligently understand and appreciate the scientific environment around us. Closely associated with scientific literacy is what is often called "scientific temper". This attribute is acquired over a long period of time and is a by product of science literacy. In the broadest sense, this means the recognition of logical consistency in ideas, the realization of cause and effect and the capacity to judge based on merits and demerits, even on issues not related to science.

2.3 THE FIRST STEP

While engaged in the task of spreading science literacy different sections of society need to be treated differently as far as the content and the method of delivery are concerned. The first step is targeted at the vast mass of uneducated and semi-literate population inhabiting our remote villages and towns. The problem here is not so much as to instill scientific knowledge among the population but to make them aware of the benefits accruing from modern day developments in science. First and foremost, they must be made aware of the evil effects resulting from following age old superstitious beliefs and practices. As Bertrand Russell rightly remarked, *"fear is the main source of superstition and one of the main sources of cruelty"*.

In our villages it is common to witness sacrifice of animals and torture of small children, all in the name of propitiating evil spirits who are blamed for causing diseases and misfortune. To counteract this, the villagers must be made aware of the curative aspects of modern medicine in eradicating, dreaded diseases like small pox, and polio, from the face of the earth. Then there are other evil practices like "Sathi" and honor killings which have no sanction of law and are based on ancient beliefs and customs. Another aspect of imparting science education in the villages and rural townships relates to hygienic practices. Open defecation common among many villagers and even among urban population is an invitation to all types of communicable diseases. Finally the young children who form the backbone of our future society should be provided with proper free education at least at the primary stage with adequate incentives.

2.4 THE CASE OF SKILLED AND SEMI-SKILLED LABOUR

Modern scientific practices demand the employment of skilled and semi-skilled labour in our factories and workshops. In addition, we have a large body of unorganized labour consisting of carpenters, blacksmiths, masons, electricians, plumbers and a host of others. Most of them are not conversant with modern work practices and are content in following age old non-scientific and inefficient methods. This large body of population should be educated on the use of modern diagnostic tools such as X-rays, ultrasound and oscilloscopes.

The mentality that if one is familiar only with one tool namely, the "hammer" then every other problem looks like as a nail, should be abolished.

2.5 EDUCATING THE BUREAUCRATS, POLITICIANS AND LEGISLATORS

In the normal developmental process, our engineers and scientists emlploy complex methodology with the view to achieving certain goals. In turn they submit these recommendations to the bureaucrats for further consideration and incorporating them in the various proposals of the government. The bureaucrats mainly do a cost benefit analysis and submit them to the politicians who may make further changes before presenting it before the legislature. In order to make meaningful decisions at various levels the decision makers involved (especially in science and technology oriented proposals) require some knowledge about the proposals under consideration. The level of such knowledge to be imparted is quite different from what is expected in the case of villagers or in the case of skilled and semi-skilled labour.

2.6 THE VOICE OF THE PEOPLE

In a participative democracy, the people's voice is paramount. Even though they may not be conversant with sufficient scientific knowledge to whet a proposal, they must at least be able to evaluate the cogency of the arguments put forward by the government. They should also understand in a broad sense, the economic, ecological and health consequences of such a proposal. Mere feasibility studies and economic viability are not sufficient to support a proposal. To cite a few examples, consider a large hydroelectric project which is planned to be executed. On the plus side the project generates clean energy at an affordable cost, but on the minus side one has also to consider the social and environmental issues associated with it before favoring a final decision. The project may involve the submersion of large tracts of land located in the catchment area with consequent damage to flora and fauna. Further, the project may displace many hundreds of families living in the areas to be submerged. The consequences of the hardships caused and the compensation to be paid to the concerned people should also be factored while evaluating the project. Similarly, in the construction of nuclear power stations, in as much as they serve as substitutes for power stations burning fossil fuels have a definite advantage. However nuclear power generation has its own drawbacks, such as safe disposal of radio active spent fuels and the probability of occurance of Fukashima type nuclear disasters. In all these cases a balanced approach taking into account a trade off between conflicting interests, have to be considered. Informed public opinion can certainly be used as an input in taking final decisions.

2.7 ROLE OF SCIENCE EDUCATION IN SCHOOLS

Science education in schools, have necessarily to cater to the needs of different sections of the population. There are some who after a short stint of studies at the primary or secondary stage may opt for taking up jobs as technicians or skilled workers. The science education in their case may be confined to a level to suit their chosen avocations. There are yet another group of students who intend to enter universities after completing their high school education. The syllabus designed to meet the science needs of such students is at present hugely unsatisfactory. The excitement and joy that science conveys, is hardly reflected in the manner in which science subjects are taught in the classroom. The syllabus does not mention anything about the philosophy of science or its history. Considering that this group provides the breeding ground to produce outstanding scientists and researchers for the country in the future, much more attention needs to be paid in structuring the science syllabus and the methodology involved in communicating the same.

Finally, it must be emphasized that science education does not cease once a student crosses the portals of the college or the university. It is a life long learning process and the joy that one derives out of it makes it worth pursuing.

Chapter 3

ORIGIN & EVOLUTION OF THE UNIVERSE – THE GRAND DESIGN

"To see a world in a grain of sand and heaven in a wild flower, hold infinity in the palm of your hand and eternity in an hour"

— *William Blake*

3.1 HISTORICAL PERSPECTIVE

Our understanding of the origin and evolution of the universe may be ranked as one of the greatest achievements of the 20th century science. It is astonishing that we the inhabitants of planet earth, a mere speck of dust in this infinitely vast universe could understand and unravel some of its deepest secrets.

Our attempt to understand the universe dates back to ancient times. As early as 3rd century BC, Plato and Aristotle proposed a model in which the earth was the centre of the universe and all the stars and planets revolved round it. This idea was fine-tuned by Ptolemy during 2nd century AD. Our concept about the universe almost remained unchanged for another 1300 years till Nicholas Copernicus proposed a radically different model known as the heliocentric model. In this model, Copernicus asserted that the sun was at the centre and all the planets revolved around it. Almost a century later Johannes Kepler and Galileo Galilee based on the meticulously recorded observations of Tycho Brahe, empirically deduced what are now known as Kepler's laws of Planetary Motion. According to these laws the planets sweep round the sun in elliptical orbits, with the sun at the focus. Thanks to the discovery of the telescope and its refinements in the hundreds of years that followed, astronomers identified millions of stars similar to the sun inhabiting the universe. At the beginning of the 20th century, the general understanding was that the universe was eternal and unchanging, occupied by a single galaxy with millions of stars inhabiting it. However, this perspective radically changed when during the 1920's US astronomer Edwin Hubble announced two path breaking discoveries. First, the fuzzy patches of light we observe in the night sky are in fact galaxies very much like our own, located at great distances from us and inhabited by millions of stars. Second, he noted that galaxies were moving apart from each other at speeds proportional to the distances separating them. This meant that the further the galaxies were from a observation point, the faster they were moving away from it. Clearly, these discoveries heralded for the first time the idea of an expanding universe. If the universe was supposed to be expanding now, then at earlier times the galaxies must have been close to each other. Extending this argument to its very limit, the universe was supposed to be born at a certain instant in time when all matter and energy coalesced into a single small region with infinitely high density and temperature. This, according to observations corroborated by computer simulation seemed to have happened about 13.8 billion years back, heralding the onset of the "Big Bang". To sum up, our present understanding is that of an expanding universe accommodating billions of galaxies with each galaxy in turn inhabited by billions of stars. Further, due to the initial momentum imparted by the Big Bang, the galaxies are supposed to move apart from each other at great speeds.

3.2 THE EVOLUTIONARY PROCESS – STEP BY STEP DESCRIPTION

The Big Bang inflationary theory described in the sequel has been recognized by cosmologists as the most credible model to explain the evolution of the universe. This model does not attempt to describe the moment of creation but rather, the growth and evolution of the universe over billions of years. According to this model, the infant universe was born about 13.8 billion years ago, out of a cauldron of radiation and elementary particles at unimaginably high density and temperature. The earliest moment after the birth of the universe that one can conceive of at about 10^{-43} seconds is known as the Planck's time. It cannot be subdivided any further into bits of time according to quantum theory. What happened between zero time and 10^{-43} seconds, is shrouded in mystery. Thereafter due to the impact of the Big Bang, the universe expanded and consequently its temperature dropped. According to hypothesis, at about 10^{-35} seconds the universe experienced an expansion of gargantuan proportions which hardly lasted 10^{-30} seconds. During this extremely small period the speed of expansion of space was supposed to have been even greater than the speed of light. The quantum fluctuation of density that occurred during this period got blown up later to astrophysical proportions because of rapid expansion. This hypothesis first proposed by US physicist Alan Guth in 1982, laid the groundwork for the structure of stars and galaxies that we see today. This entire episode in the history of the early universe is commonly referred to as the inflationary period. The inflationary energy was supposed to be responsible later for the creation of the quark soup containing nature's building blocks such as quarks, leptons and force carriers, including photons, W & Z bosons and gluons. Earlier than one microsecond, even protons and neutrons did not exist. At 10^{-5} seconds, protons and neutrons were formed from quarks - each proton and neutron comprising of 3 quarks. At about one second after the birth of the universe, the temperature and pressure of the quark-gluon soup was just right for thermonuclear reaction to take place. This lasted hardly 100 seconds, during which period, protons and neutrons interacted to form Helium, Deuterium and Lithium nuclei.

According to calculations, 25% of the mass of all atoms in the universe at that stage was in the form of Helium nuclei and the remaining 75% as protons, which incidentally formed the nucleus of the hydrogen atom. The rest of the elements of the periodic table were gradually created over billions of years in stars and also through stellar explosions. At present, the universe is made up of 5% visible matter (stars, planets, gas etc), 25% of dark matter and the remaining 70% in the form of dark energy. The explanations regarding dark matter and dark energy are discussed in subsequent sections of this chapter. During the next thousands of years that followed, the universe continued to expand and as a result temperature plummeted sharply. Prior to 100,000 years, after the birth of the universe the energy density of radiation exceeded that of matter. This prevented matter from clumping together. By about 380,000 years neutral hydrogen and helium atoms came into existence and the photons were free to start on their epic outward journey starting at about 3000 K. This radiation is now known as Cosmic Microwave Background Radiation (CMBR). This radiation by the time it reached the earth after traveling billions of kilometers got stretched to wavelengths in the vicinity of 2 to 5 mm at 2.7 K.

3.3 THE COSMIC MICROWAVE BACKGROUND RADIATION (CMBR)

CMBR was first detected by two scientists Arnold Penzias and Robert Wilson in 1964 at Bell Labs USA. While working on the development of a sensitive receiver to detect radio signals from outer space, they serendipitously discovered a background noise in all directions at microwave frequencies which they correctly identified as the remnants of radiation which bathed the universe subsequent to Big Bang. This discovery turned out to be extremely important for the following reasons.

1. It provided a virtual snapshot of the universe as it existed 380,000 years after the Big Bang.
2. It provided a strong justification for the conclusions arrived at as a consequence of the Big bang theory, including the inflationary hypothesis.
3. The very small variations of temperature of CMBR in different directions (about 0.001 per cent), reflected a slight lumpiness in the distribution of matter which subsequently led to the formation of future stars and galaxies.

Detailed measurements of CMBR were carried out subsequently in 1992 by COBE (Cosmic Background Explorer) satellite launched by NASA (National Aeronautical & Space Administration). This revealed as stated earlier small variations of the order of 1 in 100,000 in different directions.

3.4 FORMATION OF STARS AND GALAXIES

Between 380,000 years and 300 million years beginning from the Big Bang, gravity dominated and continued to amplify the density differences in the gas comprising hydrogen and helium. The first stars appeared after the end of this period. This period is referred to as the "dark ages". With no stars to light up the sky the entire universe appeared to be bleak and featureless. Star and galaxy cluster formations which came later, reached its peak 3 billion years after the birth of the universe. Thereafter as time progressed there was a gradual diminution of activities in star and galaxy formation. The next landmark as far as we the inhabitants of earth are concerned, relates to the birth of the solar system at 9 billion years. Ever since the end of inflation, the universe continued to expand at a fairly moderate rate. In the earlier periods of its expansion, it was marked by deceleration, because of gravitational influence on matter. However, a surprising phenomenon happened at about 10 billion years. The universe's expansion seemed to have shifted gear and instead of decelerating, started to accelerate. This was first noticed by two independent research teams based on their observations of distant supernova in 1998. Their findings were found to be so significant, that the U.S journal "Science", rated it as the number one discovery of 1998 in any field of research. The key roles played by the so called "dark matter" and "dark energy" in the evolution of the universe, requires further elaboration and this is the subject matter of the next section.

3.5 DARK MATTER AND DARK ENERGY

The effects of dark matter and dark energy on the growth of the universe are now well understood but the physical composition of these enigmatic entities is yet to be identified. The jury is still out on this problem and it is hoped that experiments conducted by the Large Hadron Collider (LHC) at Geneva, will yield some positive results in the near future.

Dark Matter

The existence of dark matter was first detected when orbiting stars in the periphery of our own galaxy the Milky way were found to be moving faster than it should, considering the totality of visible matter enclosed by their orbits. It appeared as if that some unknown matter which neither emits nor radiates energy and therefore invisible, but at the same time exerts gravitational pull, must be at the back of this phenomenon. This unknown entity has since been named "dark matter". Subsequent calculations show that dark matter constitutes 25% of the entire universe. Together with 5% visible matter, the total contribution of matter (visible and dark), is now estimated to account for 30% of the entire universe.

Dark Energy

Dark energy seems to be even more enigmatic than dark matter. It is held responsible for the accelerated expansion of the universe in recent times. While the gravitational energy of matter tries to rein in the expansion, dark energy does exactly the opposite. Clearly its influence is anti-gravitational in nature. Some cosmologists link dark energy with the negative energy associated with the all pervasive vacuum that fills the entire universe. They interpret that this energy with its anti-gravitational nature existed even during the early stages of the universe's evolution but was superseded by the more powerful gravitational force. However, as expansion continued, matter got diluted and its gravitational impact became less decisive. Since, the density of vacuum energy remained more or less unchanged, it overcame gravitational energy at some stage. This led to accelerated expansion at a particular stage in the life of the universe. It seems appropriate at this stage to introduce the concept of critical density which has an important bearing on the expansion of the universe. This concept is discussed in the next section.

3.6 CRITICAL DENSITY

According to cosmologists, the nature of expansion of the universe is dependant on a variable called omega (Ω), defined as the ratio of gravitational energy and kinetic energy.

While gravitational energy tends to retard the expansion, kinetic energy of matter does exactly the opposite. In the early stages of the expansion of the universe, both the energies must have cancelled each other, leading to $\Omega=1$. Had there been even a slightest deviation from equality, the one that dominated would have gained the upper hand, leading either to a big crunch or unstructured expansion of the universe. The density of matter which makes $\Omega=1$, is known as the critical density. Geometrically speaking, $\Omega=1$, corresponds to a flat universe with zero curvature and is Euclidean in nature. $\Omega>1$ corresponds to a positive curvature, analogous to the surface of a spherical ball. $\Omega<1$ corresponds to a negative curvature resembling that of a horse's saddle. However, accurate measurements have shown that the universe continues to remain flat even now with Ω more or less equal to unity. This implies that as of now gravitational energy is equal to kinetic energy. But we know for a fact that gravitational energy due to matter alone, both visible and invisible, contributes only 30% of the total energy. In other words Ω is 0.3 of critical density. The remaining 0.7 of critical density may be attributed to dark energy.

Two figures and a table appended at the end of this chapter, serves to make explicit, some of the concepts discussed. In this context, Fig 1 gives a break up of the constituents of the universe. Fig 2 is a graphical representation of the expansion of the universe and Table-1, traces in a sequential manner, the evolution of the universe from the Big Bang to the present moment.

3.7 SOME FREQUENTLY ASKED QUESTIONS

We start the section by recapitulating a few basic tenets in cosmology.

Light years, is a measure of distance and not of time. For example, 10 light years refer to the distance light has traveled over a period of 10 years. (Note, the velocity of light is fixed and is equal to 3×10^8 meters/sec).

The light or photons that strike the telescopic mirror in the laboratory as one observes a distant star, is the light that emanated from it billions of years back. Hence, what we see now is what the star was billions of years back. For all we know, the same star would have either been dead or vanished into unobservable regions of space.

The universe is so vast that what we can observe of it now is called the observable region comprising of a sphere of radius 10^{28} cms. The observed region is a sub-set of the observable region and is limited by the sophistication of the instruments at our disposal for observing the phenomenon.

When we assert that the universe is expanding, one way of visualizing it is to consider objects as stationary while the space between them is getting stretched.

The red shift/blue shift of a galaxy or a star we measure at the present moment is entirely a function of the speed at which the star or galaxy was either receding from us or advancing towards us, not now but at a time when light first emanated from it to reach us. The mathematical expression for red shift/blue shift is not a function of distance. However, one can loosely interpret by saying that the greater the red shift, the greater the distance of the object whose red shift we are measuring. With these preliminaries out of the way, we will attempt to answer some of the frequently asked questions.

The following questions are answered in this connection:

Question 1
In cosmology one talks of the space getting stretched at speeds greater than the speed of light. How is this possible, as according to our understanding, nothing can exceed the speed of light?

Answer 1
According to the theory of relativity, only the speed of objects with mass/energy is restricted by the speed of light. Space has no mass attributed to it and it is perfectly admissible for space to get stretched without any bounds on its speed.

Question 2
According to Hubble's law, all galaxies move away from each other depending upon their distance of separation. How is it that our neighboring galaxy Andromeda is moving towards the Milky Way instead of away from it, and is supposed to hit us in another 4 billion years?

Answer 2
Hubble's law manifests itself strongly while dealing with conglomerates consisting of clusters of galaxies. For lesser objects, their gravitational interaction and internal dynamics overwhelm the requirements of Hubble's law. Hence the local forces play a major role in determining how the constituent elements behave.

Question 3
The major premises on which the Big Bang theory is based on the homogeneity of the universe. But to a casual observer how can matter widely distributed in galaxies contribute to this homogeneity?

Answer 3
Considered on a small scale, the universe is certainly non-homogenous but on a larger cosmic scale which embraces within its fold entire clusters and super cluster of galaxies, large scale uniformity is still preserved.

Question 4
We do not know why there was a big bang at all or what might have existed before the Big Bang. Further we do not know whether our universe has siblings - other expanding universes far removed from our own which we cannot observe, now or any time in the future. We do not understand why the fundamental constants of nature have the values they do have at present. Are we any where near the solution of these problems?

Answer 4
The answers to all problems raised have at present no solutions. But one can never guess what the future will hold. For example, phenomena once considered far beyond the pale of our imagination are now known to exist. What is worth noting here is that cosmology, far from being a mere collection of observations a century back, has metamorphosed into a fundamental science and in the words of Sir Martin Rees – "the grandest of all environmental sciences".

3.8 A PEEP INTO THE FUTURE
It is hard to think of any intellectual activity that has changed so much during the past few decades as cosmology. Changes will continue to happen, and there are broad indications of the direction in which the universe is likely to move in future. The discovery in 1998 by two research groups, that the universe expansion far from decelerating is actually accelerating is a pointer in this direction. In this connection Dark energy is bound to have an enormous influence on the future of the universe. A possible scenario may be envisaged as follows:

As the universe expands, its observable region no doubt expands, but so does the unobservable region, the latter receding from us faster than the speed of light. Hence, over billions of years from now, we will see only a small fraction of what constitutes the total universe. Distant galaxies not bound to us by gravity will move out of our visible range, while gravity pulls nearby galaxies together. Eventually, possibly after about 100 billion years from now, all we would see will be our own galaxy enveloped in total darkness. As the Belgian priest and astronomer George Lemaitre lamented, the fire works seems to have come to an end, leaving us to wistfully recall the "vanished brilliance of a bygone age".

The relevant Figures- 1 and 2 and Table-1 follow:

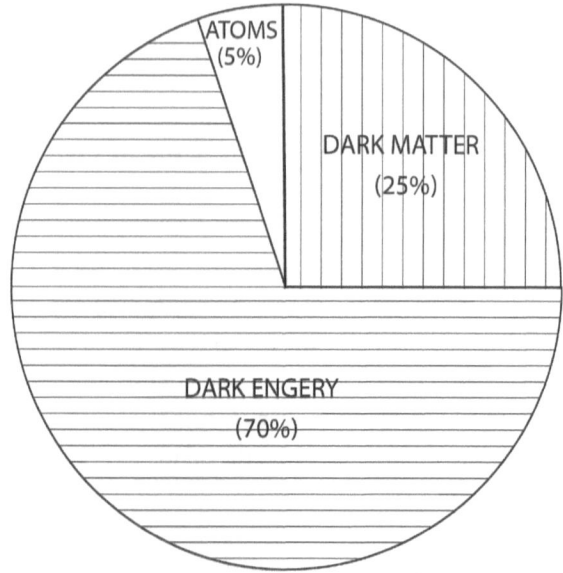

Fig 1: Break up of the constituents of the universe

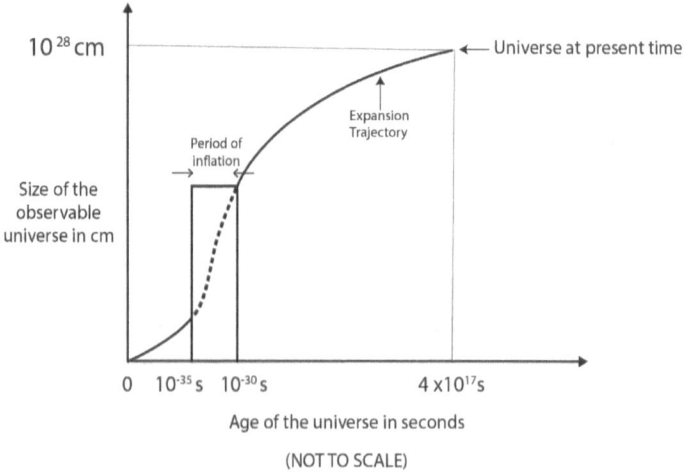

Fig 2: A graphical representation of the expansion of the universe

Table 1: Evolution of Universe from Big Bang to the Present Time

TIME	NATURE OF THE EVENT	ENERGY/TEMP 1GEV = 10^{13}K	REMARKS
Zero Time	The Big Bang	Infinite	Birth of the Universe, Quantum gravity era
10^{-43} second	Planck Time	10^{32}K	End of Quantum Gravity era
10^{-35} second	Commencement of inflation. Inflation lasted 10^{-30} second	10^{29}K	Strong force separates from non-gravitational forces
10^{-11} seconds	Electro weak force separated into weak force and Electromagnetic force	10^{15}K	All the four fundamental forces come into existence.
10^{-5} seconds	Quarks combined to form Protons and Neutrons	10^{13}K	–
1 minute	Commencement of Big Bang Nucleosynthesis (lasted three minutes)	10^{9}K	75% Hydrogen and 25% Helium formed.
380,000 years	Cosmic Microwave Background Radiation (CMBR) commences	3000K	All matter gets de-ionized
1 Billion years	Appearance of first stars and galaxies	–	First generation stars were mostly massive and "metal poor".
6 Billion years	Full fledged activity in star and galaxy formation	–	–
9 Billion years	Dark energy domination over matter commences	–	Universe expansion changes from deceleration to acceleration
9.5 Billion years	Formation of the solar system	–	Planets formed out of the rotating nebulae
13.8 Billion years	Universe as it is today	Temperature of CMBR 2.7K	–
Future	Perennial expansion/Big Crunch/Cyclical expansion?	–	Unpredictable. Most probably perennial expansion

Chapter 4

The Microscopic World of Atoms

"There is no excellent beauty that hath not some strangeness in the proportion"
— Francis Bacon

4.1 INTRODUCTION

For a complete understanding of the structure of the universe, one has to probe not only the outer space of the cosmos but also the inner space of the atom. The microscopic world of the atom deals with of sub atomic particles and their interactions. One may recall that according to the Big Bang theory, immediately after its birth, the universe was a hot dense soup of sub atomic particles at extremely high temperature and density. This heady mix was supposed to be a hundred thousand degrees hotter than the sun. These subatomic particles play a crucial role in the evolution of the universe. They also participate actively in thermonuclear reactions taking place inside stars. Hence their study is important in the cosmological context.

Way back in the 3rd century B.C the Greek philosopher Aristotle believed that all matter in the universe was made up of four basic elements- earth, air, fire and water. These elements were supposed to be acted upon by two forces namely, levity (the tendency for air and fire to rise up) and gravity (the tendency for earth and water to sink). Aristotle further believed in the continuity of matter which enabled one to divide it into smaller and smaller bits without reaching any limit. However, even earlier than this period in the 5th century B.C, a group of philosophers headed by Democritus believed that there is a limit to the division of matter and ultimately one would reach a stage where it would be impossible to divide it any further. Arguments for and against these two view points continued over centuries until 1802, when the British chemist John Dalton proposed what is known as the "law of definite proportions". Dalton postulated that the masses of various chemical elements that are required to make a given chemical compound always combine in the same ratio. This could be described in modern parlance as the coming together of atoms to form molecules. It has now been established that atoms are the smallest indivisible particles of an element which retains all the chemical properties of the element. There are 92 such stable elements in the universe as listed in the periodic table. However, with further advances in science, the inside structure of the atom itself came under scrutiny. In 1898, the British physicist J.J Thomson (who had earlier identified a negatively charged particle called the electron) suggested that atoms could be made up of positively charged lump of matter with electrons embedded in it like raisins in a fruit cake. In 1919, the New Zealand born physicist Earnest Rutherford conducted some classic experiments in order to probe deep inside the atom. These experiments indicated that atoms were made of positively charged nucleus around which electrons were orbiting, much the same way as planets were orbiting the sun. Further, the positive charge of the nucleus balanced the sum of the negative charges of the orbiting electrons so as to make the atom electrically neutral. What distinguished one element from another is the number of orbiting electrons and the corresponding equal number of protons in the nucleus. The atom was considered to be largely made up of empty space. However in 1932, James Chadwick a physicist and a colleague of Rutherford proposed that besides protons, the nucleus of any atom contained particles of zero charge with approximately the same mass as the proton.

4.2 ANOMALIES IN THE RUTHERFORD MODEL

The Rutherford model demonstrated the typical instance of the failure of classical physics as applied to microscopic matter. In his model the electron revolved in a circular orbit about a central nucleus. The electron is thus subjected to acceleration because of change of its direction, even though its speed remains constant. However, in classical electromagnetic theory, an accelerating electron radiates energy and because of this loss of energy it is expected to ultimately spiral inwards and crash into the nucleus. A second difficulty with the model was the concentration of positively charged protons packed together in close proximity within the nucleus. Here again, according to electromagnetic theory the protons should violently repel each other (like charges repel). There is no explanation for these two phenomenons in classical physics.

A solution to the first problem was provided by the Dutch physicist Niels Bohr in 1913. He postulated using quantum mechanical principles that the electron does not move in an arbitrary orbit around the nucleus of an atom and that there are some restraints imposed on its orbital path. We may recall in this context that according to wave-particle duality principle invoked by the French physicist Louis de broglie, an electron can behave both as a particle and as a wave. In the latter case, the de Broglie wave is defined by $\lambda = h/mv$, where h is the Planck's constant, m is the mass of the electron and v its speed. In order for the electron to have a stable orbit, in Niels Bohr's model, the circumference of the orbit must be integral multiples of λ. Moving in such orbits the electron does not radiate any energy as predicted in the model based on classical physics principles. The first such orbit in the Niels Bohr's model is that of the hydrogen atom having a radius a_0, given by $\lambda = 2\pi a_0$

The subsequent orbits have circumferences $2\lambda, 3\lambda, 4\lambda, ...n\lambda$. In general the radius of the nth orbit is given by $r_n = n^2 a_0$, where, $n = 1, 2, 3 ......\infty$.

For the sake of simplicity, the hydrogen atom is chosen as an example. It has one proton in its core and is orbited by one electron. The ground state of the hydrogen atom corresponds to an orbit with radius a_0 and an energy level E_1. This is the lowest energy level. Subsequent energy levels defined E_2, E_3, E_4 etc, are excited states corresponding to orbits n = 1, 2, 3, 4 ....∞.

The work needed to remove an electron from an atom in the ground state and allow it to become free is known as ionization energy. For hydrogen atom the ionization energy is 13.6 electron volts (eV). The energy level E_1 corresponding to n=1 is the lowest energy level possible and is equal to -13.6 eV. The next energy level E_2 which is higher than E_1 has a value E_2 = -3.4 eV. The values of even higher levels are E_3 = -1.51 eV, E_4 = -0.85 eV, E_5 = -0.54 eV, E infi = zero eV

We may generalize and state that $E_n = (1/n^2) E_1$ (where, E_1 and E_n are the first and nth energy states)

An electron in an atom can only occupy energy levels E_1, E_2, E_3 etc and no other level. The values of "n" associated with different energy levels are called quantum numbers. When electrons pass from one energy level to another energy level, the difference in energy level accounts for energy that is either released or absorbed. For example, if the transition is from E_i to E_j, the difference $E_i - E_j$ = h f. The frequency of photon released during this transition is given by $f = (E_i - E_j)/h$

Regarding the second anomaly noticed in the classical model of the atom, namely protons bunching together in spite of their repelling each other, is discussed in Chapter 6 while describing the forces of nature.

The structure of an atom was envisaged much before the composition of its nucleus was known. The reason is that the forces that bind the nucleus together are very much stronger than the electric forces that repel the protons in the nucleus. Changes in the structure of an atom which occur for example when photons are emitted or absorbed or when a chemical bond is formed or broken involve only very little energy (a few electron volts) as compared to the huge energies involved (of the order of millions of electron volts) when the nucleus of an atom is torn apart. In the next section we introduce certain commonly used terms to describe atomic structure.

4.3 EXPLANATION OF SOME COMMONLY USED TERMINOLOGY
Electron Volts
In atomic physics the electron volt (eV) is the commonly used unit of energy. One eV is the energy when an electron is accelerated through a potential of one volt. Since work done is charge times voltage, namely qv.

Where, q is the charge in columb and v is the voltage in volts. Thus
1 eV = 1.602 x 10^{-19} x 1 volt = 1.602 x 10^{-19} Joules

Ionization Energy

It is the work needed to remove one of the electrons from its orbit round the atom or the work needed to break apart a molecule into separate atoms. Thus the ionization energy of Hydrogen is 13.6 eV and that of Nitrogen 14.5 eV. Higher energy levels in nuclear physics are expressed as Kilo eV (10^3 eV), Mega eV (10^6 eV) and Giga eV (10^9 eV).

Atomic Number

Refers to the number of protons in the atomic nucleus. It differs from atom to atom and is denoted by the letter "Z". The value of "Z" for hydrogen is 1, for Helium 2 and for Uranium 92.

Isotopes

It turns out that in many cases different atoms of the same element have different number of neutrons in its nucleus, with the number of protons remaining the same. Such elements are known as isotopes. For example an isotope of hydrogen atom known as Deuterium has one proton and one neutron in its nucleus. Another isotope of hydrogen known as Tritium has one proton and two neutrons in its nucleus.

Atomic mass unit

This refers to the sum of the masses of protons and neutrons in the nucleus plus the mass of electrons orbiting, the atom. It is usually expressed in mass units where 12 units correspond to the mass of a carbon atom with 6 protons and 6 neutrons in the nucleus along with 6 electrons orbiting it. According to this definition 1 mass unit = 1.66054×10^{-27} Kg. The atomic mass in Kg as well as its value in terms of atomic mass unit is given in Table 1.

Table 1: Masses of particles in different units

Particle	Mass (Kg)	Mass Unit	Equivalent Mass in millions of eV
Proton	1.6726×10^{-27}	1.007276	938.28
Neutron	1.6750×10^{-27}	1.008665	939.57
Electron	9.1095×10^{-31}	5.486×10^{-4}	0.511

Atomic mass number of the nucleus

This refers to the number of nucleons in the nucleus and is denoted by the letter A. According to this definition the number of neutrons in the nucleus equals A − Z Properties of isotopes may vary drastically with mass number. For example, Tritium with mass number 3 is radioactive and unstable whereas Deuterium and hydrogen with mass numbers 2 and 1 respectively are stable and not radioactive.

Fig 1 indicates the various energy levels of the hydrogen atom starting with n=1, up to n=infinity.

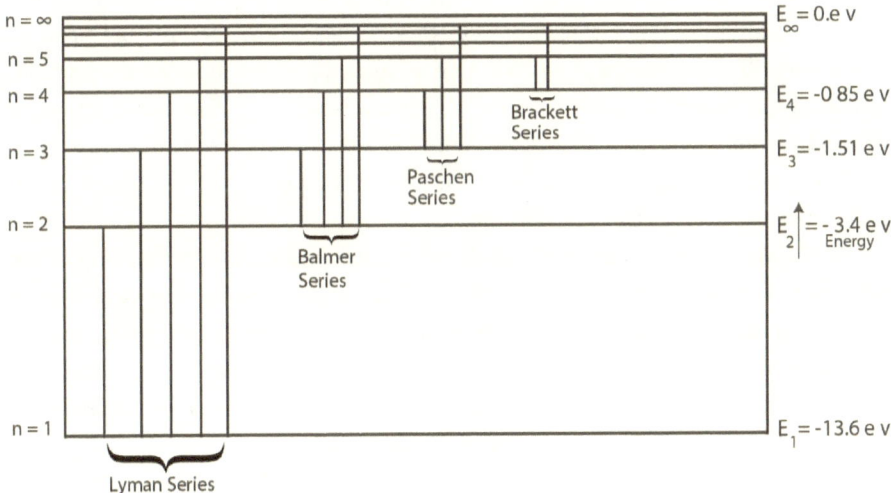

Fig 1: Energy Levels of the Hydrogen Atom

4.4 DISCOVERY OF SUB-ATOMIC PARTICLES

Up to about 1930, the only elementary particles identified were the electron, proton and the neutron. The next few decades saw the discovery of a host of particles with hardly any semblance of order amongst them. To start with, Wolfgang Pauli the Austrian born physicist postulated the existence of a particle now known as the Neutrino. These particles are extremely small, almost mass less and rarely interact with matter. Further, they carry no charge. Neutrinos are produced in prolific quantities during thermonuclear fusion inside stars. A neutrino with average energy can pass through a lead wall several thousand kilometers thick without undergoing any change in its trajectory. Every second billions of neutrinos are passing through our body without our being aware of it. They far outnumber electrons and protons in the universe by a billion to one.

In the late 1930's another elementary particle called the Muon was identified by researchers while working on cosmic rays. It was similar to the electron, but 207 times heavier. At the time of its discovery, it was considered redundant so much so that it led the exasperated particle physicist Isidor Isaac Rabi to exclaim "who ordered that". The discovery of more particles followed. In 1928 the British physicist Paul Dirac predicted the existence of a particle called the positron – a particle with the same mass as the electron but carrying positive charge. The existence of this particle was confirmed by physicist Paul Anderson of USA in 1932. This opened the doors for the existence of a class of sub-atomic particles called anti-particles which have the same mass as the particle under consideration but carry opposite charge. Around 1963, physicists made a startling discovery. Led by Murray Gellman of USA they postulated that protons and neutrons which were till then considered indivisible, were made up of sub-atomic particles called Quarks. Their name was given by Murray Gellman based on a character in the novel written by James Joyce titled "Finnegans choice". Quarks are supposed to be of two kinds- the "up quark" carrying a charge of +2/3 the charge of a proton and the "down quark" with a charge – 1/3 the charge of a proton.

At the most fundamental level, visible matter was thus seen to be made up of just two types of particles – quarks and electrons. Physicists defined the first generation of particles consisting of electrons and neutrinos collectively known as leptons and the up and down quarks collectively known as quarks. A second and third generation of particles subsequently emerged. The second generation of particles was identified as Muon (a counterpart of the electron) and the third generation was called Tau particle (a heavier counterpart of the Muon). A semblance of order among the host of elementary particles was established by arranging them in what is known as the Standard Model. The standard model of elementary particles embodies one of the most successful scientific theories ever devised. In essence it postulates that just two classes of indivisible matter particles exist namely, quarks and leptons. A mix of quarks and leptons make up the atom which is the basic building block of all visible matter. The interaction between sub-atomic particles is governed by four forces. We are familiar with two of the forces, namely gravity and electromagnetic force. The less familiar ones are the strong and weak nuclear forces. These four forces are mediated or conveyed by a set of particles known as Bosons.

A certain amount of order and symmetry was introduced among the elementary particles by the so called Standard Model of particle physics. This elegant description provides the theoretical underpinnings of particle physics. It has been experimentally verified down to a scale of 1/1000 the size of atomic nucleus, in particle accelerators.

Fig 2 gives a detailed pictorial representation of the particles associated with the standard model including their weight, their charges and their associated spins. According to this model, there are 3 generations of sub-atomic particles.

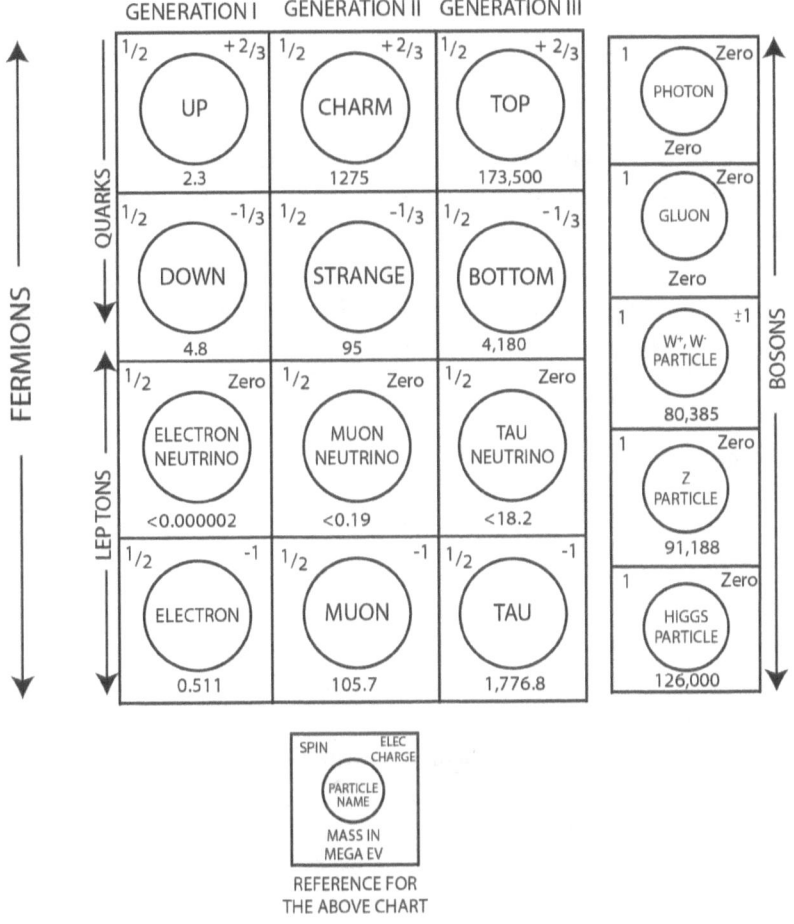

Fig 2: Standard Model of particle physics

Three generation of particles are:
- Generation I – refers to electrons, neutrons, up and down quarks.
- Generation II – refers to Muons (a heavier version of electron), Muon neutrinos, charm quark and strange quark.
- Generation III – refers to Tau particles (heavier version of Muon), Tau neutrinos, top and bottom quarks.

Besides these, there are particles which convey the fundamental forces- namely electromagnetism, strong nuclear force and weak nuclear force. These are described by the generic name Bosons.

The boson conveying electromagnetic force is called photon.

The boson conveying strong nuclear force is called Gluon.

The bosons conveying weak nuclear force are called "W" and "Z" particles.

Besides the above, we have the recently experimentally verified Higg's Boson which imparts mass to other elementary particles depending upon their interaction with the Higgs field Altogether the standard model thus describes 17 sub-atomic particles.

Major advances in particle physics interaction during the latter half of the 20th century, relate to Quantum Electro Dynamics (QED) and Quantum Chromo Dynamics (QCD). The former modified Maxwell's field equations taking relativity into account and the latter achieved the same result dealing with quarks and gluons instead of photons as is the case with QED.

Both have proved to be hugely successful theories. Later on during the 1970's, Pakistan born physicist Abdus Salam and American physicists Steven Weinberg and Sheldon Glashow, through their work unified the electromagnetic and weak forces to propound what is now known as the Electro-weak theory.

The electrons and neutrinos in generation I and the corresponding particles in generation II and III are collectively called Leptons, taking into account the fact that their masses are extremely small. The "up" and "down", "charm" and "strange" and "top" and "bottom" quarks are simply labeled "Quarks", as they have much larger masses compared to leptons. The mass and charge of each of these particles are indicated in Fig 2, which is self explanatory. Leptons and quarks taken together are known as "Fermions". They follow the Fermi-Dirac statistical laws and are subject to Pauli exclusion principle. Further, a fractional spin, specifically the number 1/2 is associated with each of these particles.

In contrast, Bosons are governed by Bose-Einstein statistics. They have integral spin like 0, 1, 2 etc. Besides the elementary particles there are also composite particles like Baryons and Mesons. Baryons are 3 quark particles (protons and neutrons).

Proton is made up of 2 "up" and 1 "down" particles (uud) and

Neutron is made up of 1 "up" and 2 "down" particles (udd)

Mesons are 2 quark particles. For example the pi-meson is made up of one up and one anti- down particle, ie (u$\bar{d}$), where anti-down particle is denoted by a bar on top of the particle symbol.

Similarly Kaon (K$^+$) is made up of one anti-strange and up particle ($\bar{s}$u).

There are several more of such particles belonging to both Baryon and Meson categories, of which only a few have been mentioned above.

Fig 3 gives the graphic structure of a hydrogen atom including the composition of proton and neutron.

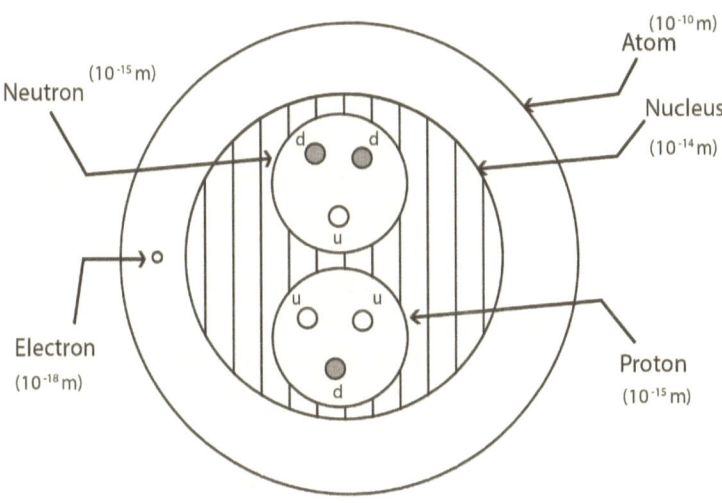

Fig 3: The inside structure of the hydrogen atom

4.5 THE HIGGS BOSON

The Higgs Boson is an elementary particle in the standard model of particle physics. Its importance in particle physics was first suspected in 1960 when Peter Higgs and a few other physicists postulated its existence. However it took over 50 years to identify it in the laboratory. In July 2012, scientists working at CERN, the European laboratory for particle physics near Geneva confirmed that they have finally sighted the Higgs particle using the Large Hadron Collider (LHC). Higgs Boson, known to the public as the "God Particle" confirmed in every respect, the original predictions of Peter Higs of Britan and Francois Engelert of Belgium. The discovery was of such fundamental importance that both the scientists were jointly awarded the Nobel Prize in physics in 2013. Higgs particle is a boson with zero charge and an equivalent mass of 126 G eV. If we examine the masses of various particles in the standard model, they cover a wide range. For example, the electron is 350,000 times lighter than the heaviest particle namely the top quark. This wide range in the mass of sub-atomic particles, have for long intrigued particle physicists. Every particle has a field associated with it and Higgs showed that the hypothetical Higgs's particle that he proposed had an associated Higgs field, According to theory, the elementary particles in the standard model interacting with this field is the cause for their acquiring appropriate masses. The stronger the interaction the greater will be the mass of the particle. For example, photon does not interact at all with the Higg's field and hence its mass is zero. Higg's particle has hitherto remained a missing link in the standard model. Its recent discovery is bound to have ramifications in the realization of the so called "Grand Unified Theory" in particle physics.

4.6 SUMMARY

The standard model continues to remain one of the most elegant theories in particle physics fully substantiated by experiments. However, it has its own limitations also. The questions often asked in this connection are:

Why do we have two types of fundamental particles, namely the leptons and the quarks, instead of a single particle to handle all situations?

After all, the standard model represents only 4% of the contents of the universe. What about the remaining 96% which includes dark matter and dark energy?

The standard model is completely silent about the gravitational force and the boson associated with it. Indeed, in the context of the large scale structure of the universe, gravitation plays a dominating role and any model which ignores it is bound to be incomplete.

The search involving hunting for structures within quarks and leptons is now brighter than ever before. With the Large Hadron Collider (LHC) attaining its full working potential in the near future, scientists are on the look out for possibly a fourth generation of quarks and leptons. The next decade is bound to witness a new foray into the sub-atomic realms leading to discovery new particles. Indeed, exciting times are ahead for the enterprising particle physicists.

Chapter 5

THE BIRTH AND DEATH OF STARS

"The spacious firmament on high with all the blue ethereal sky and spangled heavens a shining frame the great original proclaim"

– *Joseph Addison*

5.1 THE COSMIC ARCHITECTURE

On a dark and cloudless night, the sky bespangled with myriads of stars offers a magnificent and awe-inspiring spectacle. Yet, astronomers tell us that those stars that we see now are not permanent fixtures and that after a few billion years, they will all be dead and gone, replaced by new ones born out of the ashes of the old. This birth and death cycle does not continue for ever and perhaps after a hundred billion years from now, there may not be any more stars and the sky may look dark, desolate and featureless. This clinical analysis of a natural phenomena strikes different chords among different sets of people. To the philosopher, living in his own ivory tower, this is a clear affirmation of the impermanence of all worldly objects pointing out that even stars are not exempt from the irrevocable birth and death cycle. The romantic poets and artists have a different take on this. They feel that the scientists with their logical explanations supported by ugly facts have reduced the sublime and beautiful to mere matter of fact details. Be that as it may, the sheer joy of understanding the hidden secrets of nature far outweighs all of the objections voiced earlier. Clearly, knowledge promotes wonder, more effectively than ignorance.

To proceed with our narrative, it all began about 300 million years after the big bang. That was the time when the first stars and galaxies began to form out of vast clouds of gas and dust. The galaxies soon turned out to be the abode for billions of stars. As the universe expanded, neighboring galaxies tenuously bound together by gravitational forces, formed clusters of galaxies and these clusters in turn expanded to form super clusters. This process did not go on indefinitely as the continuous expansion of the universe resulted in the super clusters moving far apart, too feeble for gravitational forces to hold them in check. At this stage, the so called Hubble's law exerted itself resulting in the clusters of galaxies moving apart at great speeds proportional to the distances separating them. The formation of stars and galaxies reached their peak activity 3 billion years after the birth of the universe. Thereafter, the star formation activity came down steadily mainly because there was not enough visible matter left over to continue the process. However, new galaxy formations and merging of neighboring galaxies continued at a lesser pace. Our solar system took shape 9.5 billion years after the big bang. Eight billion years after the big bang, a strange phenomenon was noticed. The rate of expansion of the universe instead of decelerating as was generally expected started accelerating. This phenomenon is bound to have far reaching ramifications as far as the future of the universe is concerned. Having presented the big picture we shall go into a more detailed analysis in the next section.

5.2 BIRTH AND DEATH OF STARS

To begin with, it must be noted that the formation of stars and galaxies is not a one off event. It is a process rather than an event evolving over billions of years. The first star forming systems consisted mainly of ordinary matter (hydrogen atoms) along with dark matter. The denser regions of the gas containing these elements contracted into star forming clumps, each hundreds of times as massive as the sun. Some of the clumps of the gas thereafter collapsed on account of gravitational pull to form massive luminous stars.

A clump of gas should have a minimum mass to collapse under its own gravity. This mass is known as the Jeans mass, named after the British astronomer Sir James Jeans. According to his analysis, the minimum required mass to start star formation, is proportional to the square of the gas temperature. In the early days of star formation, the temperature of the universe was quite high and therefore the Jeans mass required for star formation was almost 1000 times bigger than what was required billions of years later as a result of the gradual cooling of the universe. Hence, the early stars comprising only of hydrogen and helium gas were many times more massive and luminous than our own sun. As the gas underwent contraction due to self gravity, the core temperature became hot enough to start the nuclear fusion process, which converted hydrogen into helium. The heat generated by this process created an outward push of radiation which balanced the inward pull due to gravity. This balance prevented the gas from contracting further. However, the hydrogen fuel required for thermo-nuclear fusion got quickly exhausted in the case of massive stars. With no outward pressure to oppose the crushing force of gravity pulling the gas and matter towards the core of the star, the temperature of the star became so high that the entire mass violently exploded scattering the outer contents of the star into intergalactic space, while retaining the core. Such an explosion is called the Supernova. All stars with their original mass exceeding about ten sun masses are likely to go through this process. The core left behind may still be quite massive. If the original mass of the star is no more than about five sun masses, the star ends up in what is known as a Neutron star. The neutron star comprises of at least about 1.4 solar masses compressed into a sphere of radius of about 8 Km, with a density comparable to that of the atomic nucleus. To recapitulate, we have so far dealt only with massive stars whose original mass exceeds 10 to 20 times the solar mass. More massive stars ultimately collapse into what are known as Black holes. Further details about the neutron stars and black Holes are explained in subsequent sections.

5.3 LOW MASS STARS

As stated earlier, the Jeans mass requirement for star formation of medium mass stars with mass less than 4 solar masses is quite small. The sun is a typical example of a medium mass star. In this case the thermonuclear fusion process takes place at a much slower pace and therefore the life of the star gets extended up to about ten billion years. Further, most of the medium mass stars are third generation stars and they contain besides hydrogen and helium, heavier elements such as oxygen and carbon. It may be noted here that due to problems associated with the stability of the atomic nucleus, the nuclear fusion process cannot transmute elements of lower atomic mass to those of higher atomic mass, beyond the element iron. The presence of elements with higher atomic mass found in the composition of third generation stars may mostly be attributed to supernova explosions which had occurred earlier and out of whose debris the stars were born. To continue with the narrative, when the nuclear fusion ceases, the outer mantle of the star consisting mostly of gas is forced to expand outwards and envelop the star. At this stage the star appears to look like huge ball and is called the ***Red Giant***. It takes another 5 billion years for our sun born 4.5 billion years back to reach this stage. Meanwhile the core of the star continues to shrink gradually and a stage is reached when quantum mechanical forces come into play in star formation. In particular, the so called ***electron*** and ***neutron*** degeneracy pressure exert themselves. These are dealt with in the following section.

5.4 WHITE DWARFS AND NEUTRON STARS

Two quantum mechanical principles are relevant in the evolution of medium mass stars at this stage. They relate to the Pauli's exclusion principle and the Heisenberg's principle of uncertainty. The former prohibits ***Fermions*** such as electrons, protons, and neutrons from occupying one and the same quantum state. The uncertainty principle on the other hand, states that the more accurately we know the position of a particle, the less accurate we are about its momentum. Applying the uncertainty principle to the rapidly shrinking case of the star at this stage, the electrons in the atom are forced to occupy progressively decreasing volumes of space. As their positions get localized, their corresponding momentum acting outwards increases. However, since momentum is a product of mass and velocity, it can increase only up to a limit dictated by the speed of light. Till that limit is reached, the inward acting gravitational force gets balanced by the outward momentum. After this limit is crossed, electron degeneracy pressure becomes ineffective and the gravitational force is free to further crush the core of the star. In this connection, the Indian astrophysicist Subramanyan Chandrasekar proposed a limit called the ***Chandrasekar limit***. He showed that the maximum mass that core can have to counter the electron degeneracy pressure is 1.44 solar masses. If the core mass is below this value, then a stable equilibrium is established and the resulting star is called a ***White Dwarf***. If the core mass exceeds 1.44 solar masses, then the star will shrink further. At this stage the core temperature is such that the protons combined with the electrons to form Neutrons, in a process known as *Inverse beta decay*. The neutrons are now subject to the same uncertainty principle we applied in the case of electrons and consequently the neutron degeneracy pressure comes into play to balance the gravitational pull. This results in the formation of a Neutron star. If the mass of the core exceeds a certain limit, even neutron degeneracy pressure will be unable to counter the gravitational pull and the star gets transformed into what is known as a *Black hole*.

5.5 THE MYSTERY OF THE SUPERNOVA

Supernova is a spectacular event which marks the end of a massive star. For a brief period, this phenomenon outshines billions of stars in a galaxy. They are also instrumental in seeding space with heavy elements. Supernova occurs when because of the crushing force of gravity the core of the star explodes releasing tremendous amounts of energy across the entire electromagnetic spectrum. A different kind of supernova explosion is associated with White dwarfs. Under normal circumstances, White dwarfs are stable stars and will quietly fade away. However, if the White dwarf happens to orbit another star, possibly a Red giant, it may accrete matter from its companion star and reach the Chandrasekar limit of 1.44 sun mass. This sudden acquisition of matter by the White dwarf will result in a supernova explosion which goes by the name of *Type-1(a) supernova*. The importance of Type-1(a) supernova stems from the fact that the explosion happens when the star has attained a mass of 1.44 solar mass. The characteristic glow of the debris is defined as a *Standard candle* and is used for the measurement of distances of far away galaxies with high red shift.

5.6 MORE ABOUT NEUTRON STARS, AND BLACK HOLES

As mentioned earlier, a neutron star is the collapsed core containing about 2 to 3 solar masses, left behind after a Supernova explosion. The original star mass may be about 8 to 20 solar masses. If the mass is still higher, the star ultimately collapses into a black hole. The neutron star contains about 80% neutrons, 10% electrons and the remainder is occupied by protons. The neutrons provide the outward neutron degeneracy pressure, preventing a collapse of the core thus providing stability to the neutron star. Neutron stars possess strong magnetic fields as the original magnetic field of the star is concentrated over a much smaller area (neutron stars are hardly 16 Km in diameter). The magnetic field strength may be as high as 10^{11} times that of the earth.

Pulsars

Pulsars are a special category of spinning neutron stars. It was discovered in 1967 by Jocelyn Bell, a graduate student working with Professor Anthony Hewish at Cambridge University in England. From our vantage point i.e the earth, pulsars appear to pulse light with each rotation. Their light, like that of a light house beam sweeps across the earth.

Some pulsars emit visible light, X-rays and even gamma rays. All pulsars are neutron stars, but the converse is not always true. There are neutron stars which do not radiate radio waves or X-rays in a regular and steady fashion.

Pulsars are of two types namely, isolated and binary pulsars. ***Isolated pulsars*** produce radiation primarily through their rotation and they gradually slow down and cool off. Their light is generated by electrons which are caught in the pulsars strong magnetic field concentrated near their poles.

Binary pulsars are those in orbit with a companion star, most often similar to our sun. Such a pulsar can accrete matter from its stellar companion when their orbits bring them close to each other. The violent accretion process can heat the material being transferred and produce X-rays. The X-ray light from this matter under the influence of the magnetic field can also appear to pulsate at the pulsars rotational rate. Binary pulsars are referred to as ***X-ray pulsars***.

X-ray binaries were first discovered by Riccardo Giacconi and his collaborators in early 1960's. The process of accretion can speed up the spin of a binary pulsar as the accreting material hits the pulsar at a grazing angle. Their rotation can reach hundreds of revolutions per second. Over millions of years, the gradual draining away of material from the once healthy star, can completely impoverish it and there will come a stage when no trace of the star remains. Our galaxy is supposed to contain 100,000 pulsars, of which we have hardly identified about 1000 pulsars at present.

Black Holes

Black holes are among the most exotic objects of the universe. As discussed earlier, stars associated with a few solar masses end up as **White dwarfs**. Stars containing higher mass end up as **Neutron stars** and stars with very high mass end up as **Black holes**. Astronomers consider black holes as a body exerting such intense gravitational pull that nothing, not even light can escape from its immediate vicinity. A black hole has two parts. At its core resides a singularity-an infinitesimal point into which all matter in its vicinity gets compressed. Surrounding the singularity is a region of space from which escape of matter is impossible. The boundary of this region is called the ***Event Horizon***. Once a body crosses the boundary of the event horizon, it loses all chances of ever coming out of it. Whatever light the falling body emits is also trapped inside so that an outside observer can never see it again. The body eventually crashes into the singularity. Event horizon seems much easier to comprehend than black holes. However, the singularities associated with black holes are clearly mysterious. These are places where the strength of gravity becomes infinite and the known laws of physics breaks down. Singularities invariably arise during the collapse of a giant star. Probably, quantum theory offers a way out to prevent the strength of gravity becoming truly infinite at this stage. The search is still on to develop a quantum theory of gravity to explain singularities.

The radius of the event horizon is known as the ***Schwarzschild radius*** and is proportional to the mass of the black hole. For a star with one solar mass, the radius is 3 Km, a quarter of a million times smaller than the sun's present radius. A black hole is defined by their twin parameters - its mass and its spin. The accreting gas forms a disk in the plane of the equator of the spinning black hole and jets of high energy plasma are beamed out along the opposite poles. Black holes come in various sizes. There are isolated black holes resulting from the death of massive stars with a mass of a few hundred solar masses and there are huge behemoths containing billions of solar masses. The latter known as super massive black holes, reside at the center of galaxies or in its cluster. True to their cannibalistic nature they devour stars, galaxies and black holes in the immediate vicinity of their respective event horizons. The black hole nearest to us is known as ***Sagittarius A*****, located at the center of our galaxy known as The Milky Way. It is about 4 million times the mass of the sun.

5.7 QUASI STELLAR RADIO SOURCES (QUASAR)

A Quasar, also known as ***Quasi Stellar Object*** (QSO) is an active galactic nucleus of very high luminosity. Quasars emit energy across the electromagnetic spectrum and can be observed at radio, infrared, visible, ultraviolet and X-ray wave lengths. Many astronomers believe that some quasars are the most distant objects yet detected in the universe. Quasars are associated with super massive black holes residing at the center of many galaxies. It is estimated that a quasar releases energy every second to satisfy the electrical energy needs of earth for the next billion years. Quasars are powered by the accretion of matter into the super massive black holes. The material consists of clouds of gas and dust, stars, planets and even smaller galaxies. As the material whirls around the event horizon of the super massive black hole at great speeds, it gets super heated by the magnetic field and friction. This heated material emits radiation which we observe on earth as visible light, gamma rays and X-rays etc.

There are about 200,000 known quasars in the universe. It was first detected in the 1960's. Energy from the quasars takes billions of years to reach the earth. For this reason the study of quasars can provide astronomers with information about the early beginnings of the universe.

5.8 CONCLUSION

In the evolution of the universe, we saw how stars transmuted pristine hydrogen from the Big bang into heavier atoms and when the stars died, they seeded the galaxies with basic building blocks of life namely carbon, oxygen, iron etc and out of this debris future stars are formed. The human race itself is thus made out of star dust which had at some time or other gone through the fiery baptism deep inside the stars.

Chapter 6

UNIFICATION OF THE FORCES OF NATURE – IN SEARCH OF THE HOLY GRAIL

"Simplify, Simplify"
– Henry David Thoreau

6.1 INTRODUCTION

We live in an extremely complex world. During our day to day activities we come across a variety of forces- gravitational, electrical, magnetic, centrifugal, frictional, and muscular to name only a few. Can we reduce the plethora of such forces to a few fundamental ones?. This is the question we will be addressing in the following sections. As of now, four fundamental forces namely, gravitational, electro-magnetic, and strong and weak forces have been identified. They are supposed to subsume all types of forces occuring in nature. However, identifying the forces *per se*, do not mean much unless we identify the fundamental particles on which the forces are supposed to act. In this context, it has now been established that electro-magnetic force acts only on charged particles, the strong force acts only on quarks and hadrons, the weak force acts on leptons and quarks and the gravitational force acts on all particles with mass. The fundamental particles which transmits these forces to their respective destinations are known by the generic name – **Bosons**. Each boson is associated with a specific fundamental force. Thus **Photons** convey the electro-magnetic force, the **Gluon** conveys the strong force, the W^+ W^- and Z^0 particles convey the weak force and the **Graviton** the gravitational force.

To help those who may not be familiar with some of the nomenclatures used here and for easy reference, the next section offers brief explanations.

6.2 STRUCTURE OF THE ATOM – A RECAPITULATION

Atoms are made up of a nucleus in its core, with electrons orbiting the nucleus, just like planets moving around the sun. The nucleus of an atom is composed of protons and neutrons. The number of protons indicate the specific element that the atom represents. Since the atom is neutral, the total number of protons in an atom must be same as the total number of electrons orbiting the nucleus. The word **nucleons**, refers to either a proton or a neutron. While dealing with the structure of an atom, the following symbols or letters are often used. Their significance is explained in the sequel.

A - Mass number, which denotes the number of nucleons (neutrons plus protons) in the nucleus of an atom.

Z - Atomic number denotes the number of protons in the nucleus.

N - Neutron number denotes the number of neutrons in the nucleus.

Obviously, $A = Z + N$

Atomic Mass: refers to the total mass of protons, neutrons and electrons associated with an atom. This is conveniently expressed in ***Atomic mass unit***, which is defined in terms of a carbon atom containing 6 protons, 6 neutrons and six electrons. The atomic mass of such a carbon atom is arbitrarily fixed as 12 atomic units. Thus atomic mass per unit is thus calculated to be equivalent to 1.66054×10^{-27} Kg.

Isotopes: refers to elements whose atoms have the same atomic number Z, but different neutron numbers N.

For example Hydrogen atom has one proton in the nucleus with one electron orbiting it. An isotope of Hydrogen namely ***Deuterium*** has one proton plus one neutron in the nucleus with one electron orbiting it. Another isotope of hydrogen namely ***Tritium*** has one proton plus two neutrons inside the nucleus with one electron orbiting it.

Inside an atom is mostly empty with the radius of the nucleus, ten thousandth the radius of the atom. According to quantum theory, electrons are permitted to move only along certain discrete orbits, where each orbit is associated with a certain energy level. When an electron jumps from a higher energy level to a lower energy level, the energy difference **E** is the energy acquired by the photon associated with a frequency governed by the equation $E = hf$ where **h** is the Planck's constant and **f** the frequency.

Elementary particles fall into two categories namely Leptons and Hadrons.

Leptons are light particles such as electrons and neutrinos along with their second and third generation counter parts. ***Hadrons*** are heavier particles which apart from quarks include protons and neutrons and mesons. Both leptons and hadrons are subject to weak forces. However, strong forces act only on hadrons. Three quark particles like protons and neutrons are collectively called ***Baryons***.

When one atom is transmuted to a different atom, conservation principles of energy, linear momentum and angular momentum hold good. Angular momentum is associated with what is commonly known as ***spin*** of a particle. Quarks, neutrons, protons, electrons and neutrinos are all associated with ½ spin. All elementary particles have their own individual anti- particles. ***Anti-particles*** have the same mass as the original particle but their other properties like charge and spin are different. For an anti-particle the charge, is the negative of the associated particle's charge (incase such a charge exist) and the spin is negative of the spin of the associated particle. Particles with odd half integral spin (+ ½, - ½ etc.) are referred to as ***Fermions***. They obey Pauli's exclusion principle and are governed by a statistical distribution law discovered by Fermi and Dirac known as the **Fermi-Dirac statistics.** Particles with zero or integral spin (0,-1,+1,+2 etc.) are referred to as **Bosons.** Incidentally, bosons like photons, gluons and W and Z particles have spin 1 associated with it. Bosons do not obey Pauli's exclusion principle. Instead they are governed by a statistical distribution law known as **Bose-Einstein statistics.**

For easy reference, the fundamental forces, their interactions, and particles that convey them with their properties are displayed in Tables 1, 2 and 3 respectively.

Table 1: Fundamental Forces and their Interaction

Interaction	Particles affected	Range	Relative Strength	Force Transmitting Particles	Role of Interaction
Strong	Quarks and Hadrons	10^{-15} m	1	Gluon	Holds quarks together to form nucleons. Holds nucleons together to form atomic nuclei.
Electro-Magnetic	Charged particles	Infinity	10^{-2}	Photon	Carrier of electro-magnetic force all over the universe.
Weak	Quarks and Leptons	10^{-18} m	10^{-5}	W^+ W^- and Z^0	Mediates transformation of quarks and leptons associated with radioactive decay.
Gravitational	All	Infinity	10^{-39}	Graviton*	The driving force responsible for the formation of stars and galaxies.

Graviton is yet to be experimentally detected

Table 2: Leptons and Quarks

Generation	Leptons (spin ½ particles)		Quarks (spin ½ particles)	
	Charge -1	Charge 0	Charge +2/3	Charge -1/3
I	Electron (e)	Electron neutrino (ν)	up (u)	down (d)
II	Muon (μ)	Muon neutrino (μ)	charm (c)	strange (s)
III	Tau (τ)	Tau neutrino (τ)	top (t)	bottom (b)

Table 3: Mesons and Baryons

HADRONS							
Mesons (spin 0 or 1)				Baryons (spin 1/2)			
Particles	Structure	Charge	Mass	Particles	Structure	Charge	Mass
Pion(π^+)	u d⁻	+1	0.15	Proton	ddu	+1	1
Kaon(K^+)	s⁻ u	+1	0.53	Neutron	ddu	0	1.002

Note: Masses indicated in the table are multiples of proton mass which is assumed to be unity for the purpose of comparison. There are about 140 types of mesons and 120 types of baryons. Only two representative samples in each category are listed above.

6.3 GRAVITATIONAL FORCE

Gravitational force is all pervasive and we experience it during every moment of our existense. The force is derived from what is commonly known as the ***Universal law of gravitation.*** The force is always attractive

The law can be explained as follows- consider two masses m_1 and m_2 separated by a distance d. The gravitational force of attraction between the two masses is given by

$$F = G. \{(m_1 \, m_2)/d^2\}$$

where, G known as the universal constant of gravity has a value 6.7×10^{-11} in units of $N\text{-}m^2/Kg^2$. The discovery of this law is attributed to Isaac Newton. He further showed that the gravitational force that pins us down to earth is the same force responsible for movement of planets around the sun. Thus, gravitational law is seen as a universal law irrespective of weather it operates on terrestrial or celestial objects. But Newton did not explain the genesis of gravitational force: namely why particles should attract each other the way it does. However in 1915, Albert Einstein in his celebrated paper on general relativity explained how gravitation arises out of the warping of space-time in the presence of any object endowed with mass. Its action is similar to that of a marble rolling down the depression caused by a massive body placed on the top of a tightly stretched membrane. In the classical words of the physicist J.A Wheeler *"Space tells mass how to move and mass tells space how to curve"*. We now list some of the salient features of gravitational force. First, the force is entirely attractive in nature. Second it covers the entire range starting from zero to infinity even though its strength diminishes rapidly as distances increase. Third, gravity is the weakest of all fundamental forces (see Table – I).

6.4 ELECTROMAGNETIC FORCE

This force acts only on charged particles. Like the gravitational force, it obeys the inverse square law but unlike it, the force can be either attractive or repulsive depending upon whether particles involved are carrying charges of the opposite sign or of the same sign. The law governing the force F is also known as Coulomb's law and is given by

$$F = K.\{(q_1 \, q_2)/d^2\}$$

where F is the force, q_1 and q_2 are the respective charges involved and **d** the distance between them. The constant K like the constant G in gravitational force, has the value $K = 9 \times 10^9 \, N \, m^2/(coulomb)^2$ in m-Kg-sec system of units employed. Compared with G, K has a much higher value. This explains why electro-magnetic force is very many times stronger than the gravitational force. If for example, we consider the attractive force between the proton and electron in a hydrogen atom, the electric force between them is about 10^{37} times the corresponding gravitational force. In spite of this, the reason why gravitational force preponderates over electro-magnetic force on a larger scale is because charges are subject to cancellation and hence their overall effect is on an average nullified. In contrast to this, gravitational force is always attractive and builds up to a huge value, especially while dealing with massive objects.

To begin with, it appears as though magnetic force and electric force are entirely independent entities. However, unification between the two was brought about by Ampere and Faraday. Ampere's law states that moving charges (meaning electric currents) produce a magnetic field. On the other hand Faraday's law states that a changing magnetic field creates an electric field. Indeed this forms the basis for the working of the dynamo responsible for large scale electric power generation. It is thus seen that changing magnetic field and changing electric field are some how related to each other. However it was left to Maxwell to unify the two phenomenons by formulating a compact set of equations commonly known as Maxwell's equations. Maxwell further showed that changing electric and magnetic fields creates an electro-magnetic wave which travels at a speed of 3×10^8 m/sec. In 1905 Einstein showed through his special theory of relativity that no wave can propagate at a speed greater than the velocity of light. Because of this limitation in speed, electromagnetic force cannot be conveyed instantaneously between any two points A and B, located far apart. Thus, hypothetically if the sun where to vanish suddenly from the universe at a particular instant of time, the earth will continue in its orbit round the sun for a full 8 minutes thereafter (the time taken for the light from the sun to reach the earth) as if nothing has happened, before it flies off at a tangent to its orbit.

In the next two sections, we shall deal with the remaining two forces, namely the strong and the weak interactions, whose range of influence is limited to very short distances inside the nucleus of the atom.

6.5 THE STRONG FORCE

What holds the protons of a nucleus together? In the normal course according to Coulomb's law such protons should be violently repelled, but this does not happen. The force that holds all the nucleons together in spite of the repulsive force exerted by the protons is identified as the ***strong force*** or strong interaction. In this case the electrical repulsion of the protons is overcome by the strong force which acts on both protons and neutrons alike without making a distinction of whether they carry a charge or not. In this context, the neutrons play an important role. On the one hand they contribute to the binding force of nucleons provided by the strong interaction. On the other, they help to separate the protons by occupying the intermediate space between them and to that extent, reduces the repulsive electromagnetic force acting on protons. The strong interaction also helps as a binding force acting on the constituent quarks of both protons and neutrons.

What is intriguing about the strong interaction is that the coupling strength between quarks because of the strong force grows weaker as the quarks approach each other, but increases rapidly as they stray beyond a certain limit which is roughly about 10^{-13} cm. It is analogous to a dog held in leash, where it is free to move within a certain radius but not outside it. Physicists call this property ***asymptotic freedom*** associated with strong interaction. Strong interaction is transmitted by a boson known as a gluon. It has zero charge, zero mass and is associated with integral spin (unity in this case). The theory which describes how quarks interact with each other through the exchange of gluons in a quantum mechanical setting is known as ***Quantum Chromo Dynamics (QCD)***. Its counterpart in the electromagnetic theory is known as ***Quantum Electro Dynamics (QED)***, where the interaction between charged particles (electrons and protons) in a magnetic field is through the exchange of photons.

6.6 THE WEAK FORCE

Also known as the weak interactions, next to gravitational force is the weakest of all the fundamental forces. Like strong interaction, its influence is limited to very small distances of approximately 10^{-16} cm. The strong interaction which helps to keep the nuclei together cannot account for the decay of protons and neutrons inside the nucleus. In this context weak interaction is primarily responsible in initiating what is called ***Beta decay***. It also plays a role in transforming fundamental particle like the electron neutrino to muon neutrino and tau neutrino. All radioactive decays do not involve weak interactions. However, for the sake of completeness we briefly explain 5 modes of such decay as listed below. Basically, radioactive decay occurs because the existing neutron/proton ratio is not appropriate to meet stability conditions within the nucleus. Too many neutrons compared to protons in the nucleus or alternatively too many protons compared to neutrons affect stability.

Alpha (α) Decay

This involves the emission of an Alpha particle (α) from the nucleus of an atom so as to make the nucleus left behind more stable. We may recall that an α particle which represents the nucleus of the helium atom is extremely stable. It consists of 2 protons and 2 neutrons. As a consequence of this emission the atomic number Z of the parent particle gets reduced by 2 and its atomic mass number gets reduced by 4. The new particle has thus a lower atomic number and therefore belongs to an entirely different element in the periodic table (shades of alchemy here!). Further decay can take place if the transformed element continues to be unstable.

To represent α decay in symbolic language

$$_{Z}^{A}X \longrightarrow {}_{Z-2}^{A-4}Y + {}_{2}^{4}He \text{ (alpha particle)}$$

where, X is the original element, Y the transmuted element and A, Z have the usual significance.

The reason for instability is because of the high mass of the nucleus to start with.

Beta (β) decay
In this form of decay, a neutron inside the nucleus is converted to a proton resulting in the emission of an electron and anti-neutrino. We thus have:

$$n \longrightarrow p + e + \bar{\nu}$$

where, ν represents the neutrino and $\bar{\nu}$ the anti-neutrino.

The inclusion of the elusive particle neutrino in this reaction stems from the fact that in such reactions the spin has to be conserved. Thus n with spin ½ on the left hand side is balanced on the right hand side by p with spin ½ + e with spin ½ + $\bar{\nu}$ with spin -½. Symbolically

$$_{Z}^{A}X \longrightarrow {}_{z+1}^{A}Y + e$$

Note that in β decay, since neutron is converted into proton, the number A remains the same but number Z increases by 1. β decay involves liberation of anti-neutrino. It is an elementary particle with zero charge, almost zero mass and with spin -½. Lacking in mass and not carrying a charge, neutrino hardly reacts with any matter. It can travel unimpeded through an iron wall 100 light years thick before interacting!

Reason for instability is that Nucleus has too many neutrons, relative to protons.

Gamma (γ) decay
This involves transformation of a nucleus with excess energy to one of lower energy. There is no change in the number of protons and the number of neutrons before and after the decay. Symbolically.

$$_{Z}^{A}X^{*} \longrightarrow {}_{Z}^{A}X + \gamma$$

Note that the atom with excess energy is denoted by X* to distinguish it from X.

Here, γ represents gamma rays.

Reason for instability is that Nucleus has excess energy.

Positron Emission

This decay involves the transformation of a proton in the nucleus to a neutron in the same nucleus along with the emission of positron (anti-particle of an electron), plus neutrino. We have,

$$p \longrightarrow n + e^- + \nu$$

Note that charges and spin balances on either side of the relationship. Symbolically

$$^{A}_{Z}X \longrightarrow ^{A}_{Z-1}Z + e^-$$

Reason for instability is that Nucleus has too many protons relative to the number of neutrons.

Electron Capture

In this decay, a proton inside the nucleus captures an electron outside the nucleus, thus transforming a proton into a neutron.

$$p + e \longrightarrow n + \nu$$

This results in the emission of a neutrino. Symbolically

$$^{A}_{Z}X + e \longrightarrow ^{A}_{Z-1}Y + \nu$$

Reason for instability is that Nucleus has too many protons relative to the number of neutrons.

6.7 UNIFICATION OF FORCES

The four fundamental forces explained earlier in this chapter, appear to be sufficient to govern the structure and behavior of the entire physical universe from atoms to stars and galaxies. To simplify it further attempts have been made during the past few decades to further unify the existing fundamental forces.

Success was achieved in 1967 when 3 physicists – Abdul Salam, Steven Weinberg and Sheldon Glashow proposed a theory to unite electro-magnetic force with weak interactions. Known as the **Electro weak force** their result was later on confirmed by laboratory experiments. Attempts are now on to further unify electro weak force with the strong interaction. If successful this may lead to what is hopefully named as the **Grand Unification Theory (GUT)**. Finally the grand unification of gravity with other forces if achieved will be known as **Theory of Everything (TOE)**. It is the hope of some physicists that recent advances in string theory may lead to this ultimate unification. Fig 1 gives us a block diagram representation of the entire unification concept.

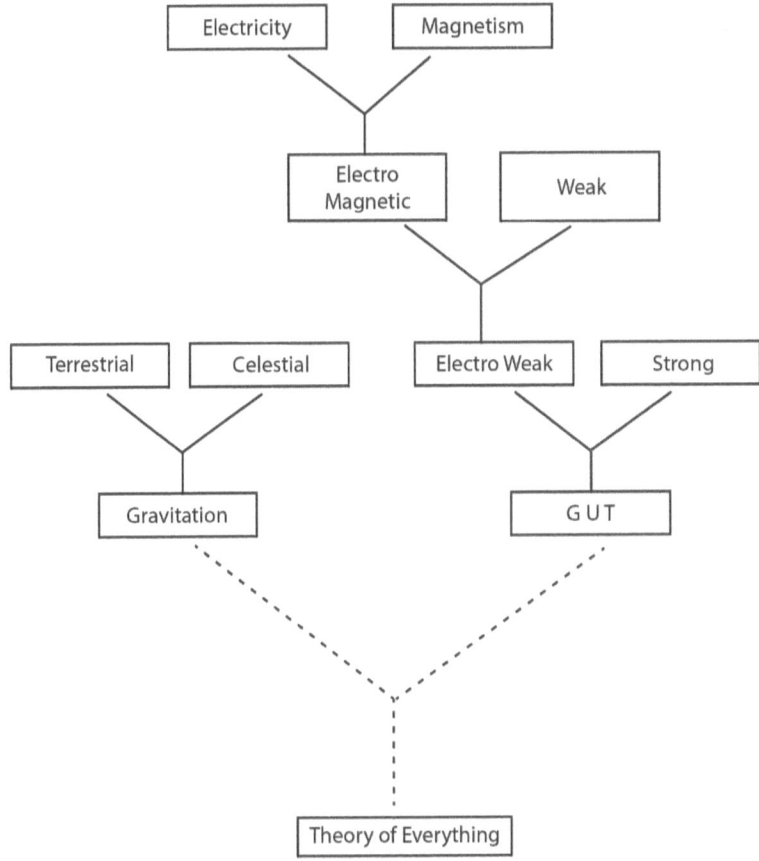

Fig 1: Unification of Forces

Chapter 7

OUR PLANET EARTH AND ITS NEIGHBOURS

"If I have seen farther, it is by standing on the shoulders of others"
— *Newton's letter to Robert Hooke*

7.1 INTRODUCTION

Our solar system comprises of the sun, 8 officially declared planets, at least 3 dwarf planets, more than 130 satellites of planets and a large number of other bodies such as comets and asteroids. The sun, which is an average star, is at the center of the solar system and all the planets revolve round it in separate orbits. The inner solar system is occupied by what are known as terrestrial or rocky planets comprising Mercury, Venus, Earth and Mars. They are called rocky because of the high density of matter contributing to their mass. All planets have two distinct motions – a rotation about their axis and motion about their orbital paths. Rocky planets are characterized by slow rotation about their axis and fast motion along their respective orbits.

Next in order of rocky planets we have the so called giant gas planets which include Jupiter, Saturn, Uranus and Neptune. They are huge compared to the rocky planets and are made up of low density matter- mostly gases like Hydrogen, Helium, and Methane. However, most of them have a hard inner core. Their speed of rotation about their axis is comparatively fast. They take a long time to complete their orbit round the sun. For example, while earth takes only one year to complete its orbit, the corresponding time for Neptune, the farthest of the planets in the solar system takes 165 earth years. The orbits of all planets are ellipses with the sun occupying one of its foci. Except, for Mercury most of them trace circular orbits. The plane in which the earth orbits is known as the *Ecliptic plane*. The orbits of most of the planets also lie on this plane. The sun's equator is contained in a plane which almost coincides with the ecliptic plane with a small departure of $7°$. All planets as seen form the top of their North poles rotate on their axis in a counter-clockwise direction (same is the case with the sun also), except for the planets Venus and Uranus. The orbital motions around the sun of all planets are also counter-clockwise. The outer solar space beyond Neptune, considered as an extension of the ecliptic plane is known as the *Kuiper Belt*. This belt also accommodates miscellaneous range of objects like dwarf planets, asteroids, comets etc. Surrounding the solar system in three dimensional space, is an imaginary sphere of huge radius (several hundreds of AU) which accommodates thousands of icy bodies and is the abode of long period comets. This space is known as the **Oort cloud**. The long period comets originating from here have highly elliptical orbits and go round the sun once in hundreds of years. What we have described so far is the big picture. The next few sections will go into more detail.

7.2 WHAT IS A PLANET?

Thanks to the recent advances in observational astronomy, it was realized that there are many orbiting bodies beyond Neptune in the Kuiper belt, some of them even bigger than Pluto that could lay a claim to be called a planet. Hence in 2006, the International Astronomical Union (IAU), had a fresh look at these bodies to decide whether they could be classified under the category of planets. In this context IAU laid down certain criteria that an orbiting body should satisfy in order to be called a planet. These are:

- A planet should orbit a star.
- It should be massive enough to settle down to a round shape.
- It should clear the neighborhood of its trajectory, of all floating objects.

The first two criteria can be easily verified, but the third one poses some difficulty. The idea is that the sum of the masses of all floating bodies within a prescribed neighborhood of the orbits should constitute only a negligible fraction of the mass of the main orbiting object. Applying this criteria they deleted Pluto from the planet category (which it enjoyed till then) and demoted it to the dwarf planet category. By the same token orbiting objects in the Kuiper belt beyond Neptune like Eris and Sedna where classified as dwarf planets. Nearer to earth, an orbiting asteroid Ceres in the asteroid belt between Mars and Jupiter was also labeled as a dwarf planet.

7.3 THE SUN AND ITS IMMEDIATE NEIGHBORHOOD

I. The Sun

The sun is the center piece of the solar system and all other celestial bodies orbit around it under the action of gravitational forces. It is an average star born about 5 billion years back and has 5 more billion years to go before it dies and transforms itself into a white dwarf. Unlike the sun, the stars which were born much earlier were massive in structure, but at the same time short lived. The various parameters of the sun *vis a vis* the first stars in the universe are highlighted in Table 1. A comparison shows the wide variation in the parameters.

Table 1: Sun versus First Stars

Parameters	Sun	First Stars
Mass	1.989 x 10^{20} Kg	100 to 1000 Sun mass
Radius	696, 000 Km	4 to 14 Sun radius
Luminosity	3.85 x 10^{23} Kw	1 million to 30 million times solar unit
Life Time	10 billion years	3 million years.

The massive nature of the sun can be judged from the fact that the sum of the masses of all the planets, their moons, asteroids etc, in the solar system accounts for hardly one half per cent of the sun mass. The suns energy output comes from fusion reaction which converts every second, 600 million tonnes of Hydrogen to 596 million tonnes of Helium. The missing mass of 4 million tonnes is transformed into energy which flows out of the sun's surface. The intense magnetic activity taking place inside the sun's core inhibits in certain regions free outward flow of heat. As a result we observe sporadic distribution of black spots on the sun's surface which are in effect low temperature regions and therefore appear black. The sun is made up of plasma (ionized gas) consisting of 75% Hydrogen, 25% Helium and a negligible percentage of heavy elements such as iron, oxygen, carbon etc. The core of the sun where fusion reaction takes place is surrounded by successive layers of photosphere, chromosphere and corona. The corona which marks the outermost atmosphere of the sun is visible to the naked eye only during a solar eclipse. At a temperature of about one million degrees Celsius, this region is even hotter than the core of the sun. The sun radiates its energy into outer space via what are known as ***solar winds***. These consist of mostly superheated atomic particles. The heating effect of solar winds can be felt as far away as the Heliosphere- an imaginary sphere of radius 10^{13} Km enveloping the entire solar system. Fortunately for us, the solar winds moving towards the earth are deflected by earth's magnetic field and to that extent its effects are minimal in our planet. Brief descriptions of the planets in the solar system follow.

ii. Mercury

It is the smallest and innermost planet in the solar system. Its orbital period around the sun of 88 days is the shortest of all planets in the solar system. Its other parameters are:

1. Diameter: 4880 Km
2. Distance from sun: 0.39 AU
3. Temperature variation: 450°C during day and -170°C during night.

It is even smaller than the Jupiter's moon Ganymede and Saturn's moon Titan in terms of diameter.

iii. Venus

It is the second nearest planet from the sun, orbiting it every 225 days. Its rotation period about its axis of 243 days is the longest compared to any planet in the solar system. It rotates in the opposite direction when compare to other planets. Venus has no satellite associated with it. Some of its other parameters are:

1. Diameter: 12104 Km.
2. Distance from the sun: 0.72 AU
3. Mean surface Temperature: 462°C

Its atmosphere consists mainly of carbon dioxide and clouds of sulphuric acid. The atmospheric pressure is roughly 90 times that of earth. Apart from the sun and the moon, it is the brightest object seen in the sky. Venus is slightly smaller than the earth. Paradoxically its surface temperature of 462°C is hotter than that of even Mercury which is closer to the sun than Venus is. Venus has no satellites and it does not have a magnetic field.

iv. Earth and its Moon

Earth is considered to be a jewel among all the planets of the sun. No other celestial body displays such a complex interplay of astronomical, chemical and biological phenomena. It supports life in all its variegated forms- vegetable, animal and human. It is the densest planet in the solar system. Some of its important parameters are:

1. Diameter: 12742 Km
2. Distance from the sun: 1 AU (i.e. 150 million Km)
3. Temperature variation: From 57°C to - 89°C

Earth has a breathable atmosphere comprising of 78% Nitrogen, 21% Oxygen and 1% Carbon-dioxide, Nitrous oxide and Methane. Two third's of the earth's surface is covered by vast oceans, which also serves as a habitat for its rich marine life. The earth has a single moon of diameter 3490 Km orbiting it at a distance of 384,000 Km. The approximate orbital period of the moon round the earth is 28 days. It turns out that the moon also rotates about its own axis once in 28 days, the same time it takes to go round the earth. As a result we on earth can see only one side of the moon and the other side known as the dark side of the moon, is permanently hidden from our view. This phenomenon is known as **Tidal locking**. This is not a peculiar phenomenon confined to earth's moon as fifteen of the moons of Saturn and eight of Jupiter are also similarly tidally locked. Unlike the earth, the moon has no atmosphere with the result that when the sun is not shinning its temperature falls to minus 163°C. The moon does not shine by its own light but reflects the rays of the sun. In this context it is far less reflective than earth, which appears 43 times brighter as seen from the moon than the moon is, as viewed from the earth. The gravitational force exerted by the moon is only 0.166 times that of the earth.

v. Mars

The fourth planet from the sun, it is also the second smallest planet in the solar system, the first being Mercury. Mars is also known as the red planet, because of the reddish dust covering its surface. It must have supported life once upon a time, but now it is completely barren. Some of its parameters are:

1. Diameter: 6780 Km
2. Distance from the sun: 1.52 AU
3. Temperature: - 145°C

Mars' atmosphere is cold and thin consisting mainly of carbon-dioxide and oxygen. Mars houses some of the largest volcanoes in the solar system. Mars has two small sized moons- Phobos and Deimos. The gravitational force exerted by Mars is only 0.38 times that of earth. We have detailed knowledge about the surface of Mars mainly due to two ambitious robotic missions of NASA. In this context two land rovers called "Spirit" and "Opportunity" landed on the surface of Mars in early 2004. The data dispatched by the two rovers confirmed that in its early formative days Mars contained water and was a much warmer and wetter planet. About 3 billion years back Mars died as a planet, bereft of all water on its surface. In September 2014, India successfully launched an orbital mission to Mars, to explore its surface and in doing so it became the fourth nation to achieve this feat.

vi. Jupiter

It is the largest planet in the solar system. It is a giant planet with a mass about one thousandth that of the sun. Jupiter and Saturn are gas giants while the other two big planets Uranus and Neptune are ice giants.

Its significant parameters are:

1. Diameter: 139,820 Km
2. Distance from the sun: 5.2 AU
3. Temperature in the clouds of Jupiter: 145°C

It takes 11.90 earth years to orbit the sun. Its gravitational force is extremely high at about 2.53 times that of earth. This huge value saves the neighboring planets such as Mars and Earth from the bombardment of asteroids in the asteroid belt, as Jupiter because of its huge gravitational force attracts them to its own surface. A large number of moons orbit this planet, of which Io (3640 Km dia), Europa (3148 Km dia), Ganymede (5280 Km dia) and Callisto (4816 Km dia) are the most prominent ones. As a matter of fact, Ganymede is the largest moon in the solar system. It is the only moon that has its own internally generated magnetic field.

vii. Saturn

It is the sixth planet from the sun and the second largest planet in the solar system after Jupiter. It is a gas planet. Its significant parameters are:

1. Diameter: 116,464 Km
2. Distance from the sun: 9.54 AU
3. Surface Temperature: -179°C to - 220°C

Saturn's mass is 95 times that of the earth. However, it is the least dense of all planets orbiting the sun. Also it has the fastest winds among all other planets in our solar system. The magnetic field of Saturn is slightly weaker than that of earth. In many ways its internal structure is similar to that of Jupiter- it generates its own internal heat. Saturn is often called a ringed planet, because it is surrounded by a series of gaseous rings. Like Jupiter, it has many moons, the biggest and best known among them being Titan (5150 Km dia). Titan was extensively explored by Cassini-Hyugens orbiter (a joint mission supported by NASA and the European Space Agency (ESA)). In January 2005, the Cassini's robotic probe Hyugens landed on Titan and according to the data sent by it, Titan's surface is covered by a dark thin frozen crust made mostly of Hydrocarbons and ice. Titan has a 600 Km deep atmosphere comprising of Nitrogen and Methane.

viii. Uranus

It is the seventh planet from the sun. It has the third largest radius among planets and fourth largest planetary mass in the solar system. Uranus does not have a rocky core like Jupiter and Saturn. Instead its material is more or less uniformly distributed. Its atmosphere comprises about 83% Hydrogen, 15% Helium, and 2% Methane. Some of its important parameters are:

1. Diameter: 50724 Km.
2. Distance from sun: 19.2 AU
3. Temperature: = 216°C

Unlike most other planets which spin about an axis nearly perpendicular to the plane of the ecliptic, Uranus's spin axis lies along the ecliptic plane. Uranus is primarily composed of rock and ice with only 15% Hydrogen and a little Helium in contrast to Jupiter and Saturn which are composed mainly of Hydrogen. Uranus has over 27 moons, most of them of small diameter and negligible mass. The gravitational force exerted by Uranus is about 0.91 times the earth's gravitational force. Most of our information about Uranus and Neptune came from NASA's Voyager-2 space craft, which flew past it.

ix. Neptune

It is the eight and the farthest known planet of the solar system. It is the fourth largest planet by diameter and the densest giant planet. Its principal parameters are:

1. Diameter: 49324 Km
2. Distance from sun: 30 AU
3. Temperature: Average temperature of -214°C

Discovered in 1846, it is the only planet whose existence was predicted earlier than the date it was discovered. This prediction was based on measurements of perturbations induced by it in the orbital trajectories of Uranus and other planets. Methane is the predominant gas in Neptune's atmosphere apart from Ammonia and Helium. Neptune has over 13 moons, the largest of them being Triton (2706 Km dia). It is the only large moon in our solar system that orbits in the opposite direction of its parent planets rotation (a retrograde orbit). Like our moon it is locked in synchronous rotation with Neptune- i.e. one side of it faces planet Neptune at all times. The moons of Saturn, Uranus and Neptune are smaller compared to Jupiter's moons. The gravitational force exerted by Neptune is similar to that of earth (about 1.14 times earth's gravity).

Table 2 which follows presents a comprehensive view of the various parameters associated with the eight planets. A few of these parameters have already been highlighted in the main text.

Table 2: Essential Parameters of Planets in the Solar System

Planet	Distance from the Sun in AU*	Rotational period	Orbital period. Earth = 1 year	Comparative Mass. Earth = 1	Eccentricity	Radius (Km)	Inclination of Spinning axis (Deg)#	No. of Moons	Gravitational Force Earth = 1g
Mercury	0.39	59 days	0.24 yr	0.06	0.21	2440	0.0	0	0.379g
Venus	0.72	243 days	0.81 yr	0.81	0.01	6052	177	0	0.91g
Earth	1.00	24 hrs	1.00 yr	1.00	0.02	6371	23.5	1	1.00g
Mars	1.52	24.6 hrs	1.83 yr	0.11	0.09	3390	25.0	2	0.38g
Jupiter	5.20	9.9 hrs	11.90 yr	318	0.05	69,911	3.0	63	2.53g
Saturn	9.54	10.7 hrs	29.5 yr	95	0.06	58,232	27.0	62	1.07g
Uranus	19.20	17.2 hrs	84.00 yr	14.50	0.05	25,362	98.0	27	0.91g
Neptune	30.00	16.0 hrs	165.00 yr	17.10	0.01	24,622	28.0	13	1.14g

Note: *AU stands for astronomical units. i.e the average distance from the earth to the sun.
1 AU = 150 Million Kilo meters
Inclination of spinning axis is with respect to the normal to the ecliptic plane.

7.4 DWARF PLANETS, ASTEROIDS AND COMETS

Pluto is a dwarf planet (2374 Km dia) in the Kuiper belt, which houses a number of bodies beyond Neptune's orbit. Its distance from the sun is about 39.5 AU. It traces a largely elliptical orbit about the sun and takes 249 years to complete one orbit. It has 5 moons, Charon, Hydra, Nix, Styx and Kerberos of which Charon is the largest among them, the rest of the moons being very small. Charon with a mean diameter of 1212 Km is about half as big as Pluto, its parent planet. The surfaces of Charon and Pluto always face each other, a phenomenon known as tidal locking. Charon orbits Pluto every 6.4 earth days and coincides with the complete rotation of Pluto about its spinning axis, namely 6.4 earth days. Pluto's rotation is retrograde, as it rotates from east to west, similar to Uranus and Venus, which also exhibit retrograde rotations.

Eris and Sedna

Eris is the second largest dwarf planet known in the solar system. Its orbit does not lie in the ecliptic plane and extends far beyond the Kuiper belt. Eris like Pluto is smaller than the earth's moon. It was discovered as late as 2005. Its mean radius is 1163 Km and is located at a distance of 96.4 AU from the sun. It has a single moon Dysnomia orbiting it. The orbital period of Eris is 561 earth years. A day on Eris lasts 25.9 hours.

Sedna is a dwarf planet in the far outer reaches of the solar system. It was discovered in 2003. At about 1000 Km in diameter, it is half the size of Pluto (2374 Km dia) and roughly the same size as Pluto's moon Charon. Sedna is located in the Oort cloud region inhabited by many icy objects and believed to be the home of many comets. It takes a very long time for Sedna to orbit the sun, about 11,400 years. Its orbit is hugely eccentric (0.854 eccentricity). Sedna has no known moons. Its temperature on a hot day is about 237.6°C.

Ceres, a dwarf planet is the largest orbiting object in the asteroid belt, spanning the space between the orbits of Mars and Jupiter. NASA's Dawn spacecraft launched in September 2007, orbited Ceres in March 2015 and made extensive studies of its surface structure. It is the only dwarf planet located in the inner solar system. Ceres was known as an asteroid for many years before it was classified as a dwarf planet in 2006. It has no moons. Other relevant parameters are:

1. Diameter: 950 Km
2. Distance from Sun: 2.8 AU
3. It completes one rotation about its axis every 9 hours (making its day, one of the shortest in the solar system)
4. Its orbital period round the sun: 4.6 years

Its axis of rotation is tilted 4° with respect to the normal to the plane of its orbit round the sun. That means it spins nearly perfectly upright and does not experience seasons like other more tilted planets do. Ceres has a very thin atmosphere and there is evidence that it contains water vapor. The day time temperature of Ceres is -73°C and the night temperature -143°C.

Brown Dwarfs

Brown dwarfs are often referred to as failed stars. They belong to neither a star nor a planet category, but lie somewhere in between. These stars shines by their own light compared to planets that reflects the light of the sun they orbit. They are massive enough to undergo stable hydrogen fusion inside their interiors over a sustained period. They are formed out of the collapsing clouds of inter stellar gas as in the case of planets. Planets on the other hand are too cold and small in size to initiate fusion. Straddling the mass range between planets and stars and sharing many of the characteristics of both, are what are known as Brown dwarfs. Their masses range between 12 to 75 Jupiter masses. This mass range is however too small to attain hydrogen fusion, but are heavy enough to fuse into Deuterium – an isotope of hydrogen. Brown dwarfs shine like feeble stars but eventually cool down to resemble planets. Astronomers have identified hundreds of brown dwarfs in the solar system and in the Milky Way. Many of them float through space as stars do and few go into orbits as planets do. They are potential contenders for harboring extra terrestrial life as they satisfy some of the criteria required to support life in habitable zones.

Comets

Comets are balls of dirt and ice (made out of water), Ammonia and Methane. Swarms of such objects roam in the Oort cloud- the vast spherical space extending half way to the nearest stars. They trace hugely elongated elliptical orbits round the sun and spend most of their orbital time away from it. Like the planets, comets do not shine on their own, but reflect sun light falling on them. The center part of the comet called the nucleus is made up of frozen gas, ice and rock. When the comet approaches the sun, its nucleus becomes warmer and as the comet melts, a tail is formed out of the water vapor. When the comet is far away from the sun, it has no tail. Comets are of various types. Long period comets like, Hale-Bopp has an orbital period of over 200 years. The short period ones have periods ranging from 20 to 200 years. In olden days the appearance of a comet invoked fear and awe among people as it was supposed to spell doom and disaster for kings and kingdoms, whenever it was sighted. However, as we know now, comets appearance is purely dictated by celestial mechanics (76 years periodic appearance, as in the case of Halley's Comet) and has nothing to do with fortunes of the country's rulers.

Asteroids

These are large masses of rock and metal. Most of these objects orbit the sun. They lie within a belt called asteroid belt. Some of them are as big as 20 Km across. It was one such asteroid that hit the earth about 65 million years back, resulting in the extinction of Dinosaurs. Asteroids do pose a potential threat to living beings on earth even though the probability of such an occurrence is quite small. Apart from asteroids the space is awash with meteoroids which are chunks of rock or bits of shattered asteroids. They continue to bombard the earth, piercing through the atmosphere and in the process often get burned emitting streaks of light.

7.5 EXOPLANETS

Planets which are outside our solar system are known as Exoplanets or extra solar planets. Most of them orbit the stars. However, there are some rouge planets which are not attached to any star in particular. With the development of increasingly powerful telescopes, scientists have been exploring regions far beyond the solar system, in search of planets orbiting different stars. The main motivation for the search is to discover alien life and habitable spaces outside the solar system. The first exoplanet was discovered in 1995. However, the modern era of planet hunting began with the launch of NASA's Kepler space telescope in 2009, specifically designed to explore such planets. Since then, over 3000 exoplanets have been discovered and catalogued, but none of them so far satisfy all the criteria required for extra-terrestrial life to exist. The most important criteria for life to thrive on a planet are:

1. The planet should orbit at a distance from the star where liquid water can exist on the surface of the planet. This is a non-negotiable condition.
2. The temperature in the planet should not swing from extreme heat to extreme cold, both of which are inimical for existence of life.
3. There must be quantities of Hydrogen, Oxygen, Carbon and Carbon-dioxide available in the planets atmosphere and on its surface which can support life.

Broadly speaking, we are looking for earth like planets orbiting sun like stars. The detection of exoplanets far removed from the solar system is extremely difficult and poses a challenge to astronomers. The procedure adopted is as follows. It must be borne in mind that reflected light from a far away planet (which reflects light from the parent star), is extremely weak and cannot be measured. However, as the planet makes a transit across the face of the star, a small strip of star light shines through the planets atmospheric gases and chemicals. As a result different wave lengths of light are absorbed by the planets atmosphere. By looking at the missing wave lengths in the stars radiation, as measured on earth, one can deduce information about the planets atmosphere and nature of its surface. This is an indirect method and the more direct method would involve observation of the planets directly using powerful telescopes. Astronomers hope that with the introduction of the proposed James Webb telescope with a much higher resolution than that of its predecessor the Hubble telescope, it may be possible to directly observe some of the exoplanets.

Chapter 8

SCANNING THE UNIVERSE

"When you cannot express it in numbers your knowledge is of a meager and unsatisfactory kind"

– Lord Kelvin

8.1 AN OVERVIEW

If we want to fully understand any phenomenon, we must be capable of measuring it. According to the astrophysicist, Martin Rees, cosmology is a subject in which observation is king. From the dawn of civilization, to just about 400 years back, all we knew about the universe came through observations by the naked eye. However, Galileo (1564-1642) was the first astronomer in 1610, to point the telescope towards the night sky in a meaningful way, thus revolutionizing our knowledge of the universe.

Our main information about the celestial bodies comes from the Electro-Magnetic (EM) radiation emitted by these bodies over a wide spectrum of frequencies. These range from radio waves at low frequency end of the spectrum to the Gamma rays at extremely high frequencies.

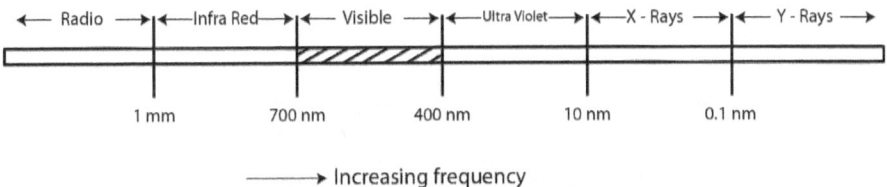

Fig 1: Electromagnetic Radiation over the entire frequency Spectrum

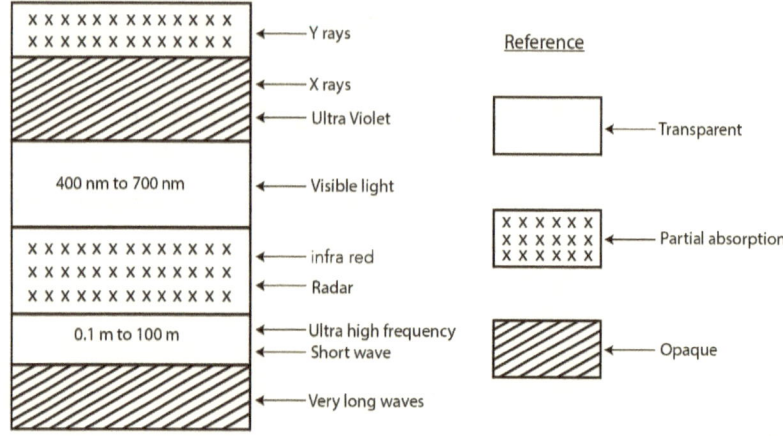

Fig 2: Filtering of EM waves by earth's atmosphere

Figure-1 attempts to give a graphical representation of frequencies across the EM spectrum. However, almost all of the EM radiation emitted, except for a narrow band associated with visible light and certain range of radio frequencies is blocked by the atmosphere surrounding the earth. Fig 2 gives information about the range of frequencies which are thus completely blocked, partially blocked or transparent with respect to earth's atmosphere. Before World War-II (1939-1945), telescopes were the only device used to scan the universe. However, during the war, the extensive research carried out on Radar indirectly gave a fillip to Radio Astronomy, which deals with observing celestial objects at radio frequencies. A direct fall out from this development was the path breaking discovery of Cosmic Micro wave Background Radiation (CMBR) in 1965 by Arno Penzias and Robert Wilson. Another noteworthy development during the 1960's was the mastering of orbiting satellite technology which culminated in the first landing of man on the moon in 1969. As a result of these developments, National Aeronautical Space Administration (NASA) of USA was able to launch Hubble Space Telescope (1990), which is an optical telescope orbiting the earth at an altitude of 569 Km. After a period of 27 years since its launch, the Hubble telescope is still in operation. This development was followed by three other great satellite borne telescopes launched by NASA namely; the Compton Gamma Ray observatory (1991), the Chandra X-ray telescope (1999) and the Spitzer Space telescope (2003), each specializing in observing different parts of the EM spectrum. Simultaneously, remarkable progress was registered in developing large ground based optical telescopes such as the twin 10m Keck telescopes located in Mauna Kea in Hawaii and also the Very Large Telescope (VLT) operated by the European Southern observatory in Atacama Desert in northern Chile. The VLT consists of 4 individual telescopes each with a mirror of 8.2m diameter which can be used either separately or in unison, to achieve high angular resolution. It scans wave lengths corresponding to 300 nm to 20 μm (visible, near and mid infra red ranges). Another major development using a different technology was in the area of Radio Astronomy, with very large radio antennae's commonly referred to as radio telescopes which were either used in isolation or by linking together multiple telescopes. This method virtually increased the aperture size of the composite telescope using a procedure known as aperture synthesis. Radio astronomy over the past few decades has led to substantial increase in our knowledge of the universe. Two other developments, though they are still in a nascent stage of development are worthy of mention. The first relates to the study of gravitational waves which emanated immediately after the Big Bang or from other cataclysmic events occuring during the collision of galaxies. The second development relates to the study of neutrinos- the elusive particles with almost zero mass, which hardly interact with matter and therefore capable of traveling unimpeded through space right from the beginning of the universe.

8.2 OBSERVATIONAL ASTRONOMY – SOME IMPORTANT MILESTONES

Astronomical observations broadly fall under two categories, namely ground based and satellite based observations. One of the main disadvantages of the ground based optical telescopes is that even in cases where signals penetrate the atmosphere and reach the observer, they suffer a certain amount of distortion because of atmospheric disturbances. This is eliminated when the observations are conducted using satellite mounted instruments, located high above the earth's atmosphere. However, on the plus side, land based instruments do not suffer from weight restrictions and are amenable to easy access for repair and modernization. For example, the twin Keck telescopes located in Hawaii each have a 10m aperture which no satellite borne telescopes possesses at present. Further, a still bigger earth bound optical telescope with a 30m diameter mirror is now being designed, which when installed will measure radiation at near ultra violet, visible and infra red frequencies corresponding to wave lengths of 0.31 μ m to 28 μm. The resolving power of such an instrument will far exceed any of the present satellite mounted optical telescopes. We shall hereafter confine our attention to describing some of the more important measuring devices currently employed in observational astronomy.

i. The Hubble Telescope

Named after the 20th century astronomer Edwin Hubble, the Hubble Space telescope was launched by NASA in 1990. After 27 years of uninterrupted service, it is likely to be replaced soon by the more modern next generation James Webb Telescope. Hubble telescope orbits the earth at an altitude of 569 Km with an orbital period of 96 minutes. With a 2.4m diameter mirror, Hubble observes the near-ultra violet, visible and near infra red spectra. As it orbits beyond the distortions of the earth's atmosphere, it is able to produce clear, extremely high resolution images as compared to ground based optical telescopes. Its observations have led to some of the major breakthroughs in astrophysics such as the accurate determination of the rate of expansion of the universe. After its launch in 1990, its instruments have been upgraded a few times with the help of space shuttle missions. In its eventful tenure, Hubble has recorded some of the most detailed visible light images ever, allowing a deeper view of the universe both in space and time.

ii. James Webb Telescope

The James Webb Space Telescope (JWST) is supposed to be the scientific successor of the Hubble space telescope, but not exactly its replacement because their capabilities are quite different. JWST will have the ability to see high red shift objects. It is now under construction and is scheduled to be launched in 2021. It has an aperture diameter of 6.5m as compared to 2.4m of the Hubble telescope. JWST will offer unprecedented resolution and sensitivity from orange red wave lengths in the visible spectrum up to infra red wave lengths corresponding to wave lengths 0.6µm to 27µm. It is supposed to be located near the Earth-Sun L2 point (Lagrange point) and will be orbiting the sun at a distance of 151.5 million Km. However, because of its special location (L2 point), its orbital period will be the same as that of the earth, even though it is located farther away from the Sun than the earth is, by about 1.5 million Km. JWST's versatile capability coupled with its high vantage point will enable it to conduct broad investigations across many fields of astronomy and cosmology.

iii. Compton Gamma Ray Observatory

Compton Gamma Ray Observatory (CGRO) was launched in April 1991 to cover X-ray and gamma ray wave lengths corresponding to 40 Pico (10^{-12} m) and 60 atto (10^{-18} m). It was deployed in the low earth orbit at 450 Km to avoid the Van Allen radiation belt. The observatory is named after the Nobel laureate Arthur Holly Compton. However, after a few years of successful operation it was deliberately de-orbited in June 2000 because of the failure of three of its gyroscopes.

iv. Chandra X- Ray Observatory

Chandra X-ray observatory (CXO) was launched in July 1999. It is sensitive to X-rays hundred times fainter than what is measured by any previous X-ray telescope. Chandra is a geocentric satellite with a highly elliptical 64 hour orbit and its mission is still ongoing as on 2017. Since earth's atmosphere absorbs the vast majority of X-rays, no earth based telescope can detect these rays. This deficiency is sought to be rectified by the installation of the Chandra space telescope. Its main aperture has a diameter of 1.2m and covers X-ray wave lengths from 0.12 nano meters to 12 nano meters. The telescope is named after the Indian born astrophysicist and Nobel laureate Subramanyan Chandrasekhar.

v. The Spitzer Space Telescope

It is a cryogenically cooled infra red telescope launched in August 2003 and turns out to be the fourth and final telescope of NASA's Great Observatories Program. Unlike Hubble, it orbits the sun instead of the earth. It has an almost circular orbit around the Sun at a distance of about 1.015 AU. With an aperture diameter of 0.85m, it is intended to cover wave lengths in the infra red range of the electro-magnetic spectrum, ranging from 3.6μm to 160μm. Its purpose is to study objects deep within our solar system and beyond. It was partially disabled in 2009 when the liquid helium supply which cools the telescope to very low temperatures was exhausted. However it continues to function with reduced capability even now.

vi. Kepler Space Craft Astronomical Observatory

This space laboratory was launched by NASA in March 2009. Named after the astronomer Johannes Kepler, the space craft was launched into an earth trailing helio-centric orbit. It is designed to survey Earth- size planets orbiting other Sun like stars. It has an approximately circular orbit around the Sun at a distance of 1 AU, with an orbital period of 373 days. The aperture diameter of the telescope is 0.95m and operates at wave lengths ranging from 480 nano meters to 890 nano meters. Two of the four reaction wheels of the telescope used for pointing the space craft towards the stars failed by May 2013. It has hence become partially disabled, but continues to function in a restricted mode. As of Jan 2015, Kepler with its follow up observations have found over 1000 confirmed exoplanets orbiting about 440 star systems.

vii. International Space Station

International Space Station (ISS) is a habitable artificial satellite station in a low earth orbit of 400 Km. It serves as a home where astronauts live and conduct experiments. Launched into orbit in 1998, the ISS is the largest man made body in low earth orbit. It normally carries a full complement of 6 crew members. ISS provides a micro-gravity environment which enables the crew to conduct experiments in biology, physics, astronomy, meteorology and a host of other fields. Earlier, the US space shuttle was used to transport astronauts and cargo to the space station, but since 2011, the shuttle program has been discontinued. The Russian Soyuz rockets are now the sole providers of transport for the crew, to and fro from the earth station.

viii. Radio Astronomy

It relates to a sub-field of astronomy that studies celestial objects at radio frequencies. There are different sources of radio emission in the universe. Apart from stars and galaxies, they include radio galaxies, quasars and pulsars. The discovery of Cosmic Microwave Back ground Radiation (CMBR) regarded as strong evidence for the Big Bang theory, was made through the use of radio astronomy. It involves using larger radio antennas referred to as radio telescopes, either singly or in groups linked together using the technique of aperture synthesis. This is achieved through mathematical signal processing technique which combines separate signals to create a high resolution image. At radio frequency wave lengths, image resolution of a few micro arc-seconds can be achieved using this technique.

8.3 THE LARGE HADRON COLLIDER

The Large Hadron Collider (LHC) is the world's largest and most powerful particle accelerator now in operation. It was built by the European organization for nuclear research (CERN) between 1998 and 2008 in collaboration with thousands of scientists and engineers from all over the world as well as hundreds of universities and laboratories. It is installed in a circular tunnel, 27 Km in circumference, at a depth of 175 meters at some places, beneath the France- Switzerland border near Geneva, Switzerland.

LHC allows scientists to reproduce conditions that prevailed in the universe, a millionth of a second after the Big Bang. LHC is exactly what its name suggests- a large collider of hadrons which are particles made up of quarks. Inside the accelerator, two high energy proton beams travel at close to the speed of light in opposite directions before they are made to collide. The beams travel in separate pipes kept at ultra high vacuum condition and they are guided around the accelerator ring by strong magnetic fields maintained by super conducting electro-magnets. The proton beams are made to collide at four designated locations around the accelerator ring, to enable scientists to make relevant measurements. The particles involved are so tiny that the precision involved in making them collide is equivalent to firing two needles 10 Km apart, such that they meet each other at the half way mark. One of the most exciting discoveries made so far while operating the LHC relates to is the identification of the mythical and elusive "God Particle" called the **Higgs boson**. This particle discovered experimentally in 2012, is the final missing link in what is commonly referred to as the Standard model of particle physics. Scientists are constantly searching through their experiments for newer particles that do not fit into the standard model, such as anti-matter, dark matter and the sub atomic particles predicted by super symmetric theories. The LHC is designed to ultimately reach collision energy level of 16 Tera electron Volts (TeV). Presently, it is working at a reduced energy of 13 TeV, comprising of 6.5 TeV per beam. Although the LHC can detect a lot of microscopic phenomena, it still leaves many questions unanswered. For example, it accounts for only 5% of the universe, the rest being attributed to dark matter and dark energy, which are yet to be identified in LHC experiments. Recently, the LHC has announced the discovery of a state of matter resembling quark-gluon plasma, possibly a state of matter that existed, a few millionths of second after the Big Bang.

8.4 GRAVITATIONAL WAVES AND NEUTRINO ASTRONOMY

What we have discussed so far relate to Electro-magnetic (EM) waves which had its origin only 400 million years after the Big Bang. Much has been achieved by studying these waves. However in cosmology, it is equally important to know what happened to the universe a few micro seconds after the Big Bang. Gravitational waves and neutrinos partially provide us the answer to such questions. A major draw back in this context is that the associated signals are extremely weak and technology is just catching up to improvise sufficiently sensitive instruments to accurately measure these signals.

i. Gravitational waves

These are ripples produced in space time which propagate as waves at the speed of light. Such waves were predicted by Einstein in his general theory of relativity (1916). Gravitational waves transport energy as gravitational radiation in a manner similar to EM radiation. Gravitational wave astronomy is an emerging branch of observational astronomy which uses space time medium to collect information about sources which produce these waves. These include binary stars, black holes, supernova and most importantly about the universe immediately after the Big Bang. To measure these waves the advanced Laser Interferometer Gravitational wave Observatory (LIGO) was developed by the US in 2008. It operates at wave lengths ranging from 43 Km to 10000 Km. LIGO operates two gravitational observatories in unison, separated by a distance of 3000 Km. This distance, corresponds(at the speed of light), for a gravitational wave arrival time difference measured by the two stations of about 10 milli seconds. This difference in arrival time is used to locate the source of the wave. After decades of search for ripples in space time (predicted by Einstein 100 years ago), scientists using the LIGO have identified gravitational waves emanating from two merging black holes 1.3 billion light years away from us, in February 2016. Indirect evidence regarding the existence of gravitational waves was available much earlier using the measured decay in orbital period of a binary pulsar by two astrophysicists Russel Hulse and Joseph Taylor in 1974. For their outstanding work, they were jointly awarded the Nobel Prize in 1993.

ii. Neutrino Astronomy

During the past few decades, a new type of astronomy has emerged. Instead of looking up into the night sky, in this case the observatories are buried deep underground, under water or under ice, not in search of photons but in search of particles called neutrinos. The neutrinos are nearly mass less and interact weakly with matter. At present several large neutrino detectors are being constructed. It is hoped that the detection of these particles emanating from extra galactic sources will enable us to study these sources and provide new information about the fundamental properties of matter.

The weak interaction of neutrinos with matter implies that they are extremely difficult to detect. This requires construction of detectors weighing several kilotons of detecting medium. In addition to being very massive, all detectors are buried deep underground. The reason is as follows. Earth's surface is bombarded by a large number of high energy particles. These particles are mostly associated with cosmic rays, gamma rays etc. Penetration of such high energy particles to reach the detector may result in interactions that mimic neutrino interactions. Burial of neutrino detectors deep underground will enable to filter out these spurious background interactions. Unlike such particles, only neutrinos can penetrate deep into the earth. The Sun which undergoes nuclear fusion is a source of abundant supply of neutrinos- (some 100 billion neutrinos per square centimeter per second). Neutrinos come in 3 types of "flavors"- the electron type, the muon type and the tau type. The neutrinos produced by the Sun during fusion are all of the electron type, but they change their flavor to muon type and tau type as they journey towards earth. This conversion of flavors is known as "oscillations" and partly explains why neutrinos possess mass even though they are of a negligible order. Oscillations are also observed in neutrinos produced in nuclear reactors. Several high energy neutrino telescopes are now under construction deep underground, under water or under ice in the polar ice cap region. These include The Super- Kamiokande neutrino laboratory in Japan, the Sudbury nuclear laboratory in Canada and the Ice-cube neutrino observatory in South Pole. An India based Neutrino observatory (INO) – a world class laboratory belonging to this category, has been in the planning stage for quite some time, but it is yet to take shape. Finally, it is expected that the neutrino telescopes may contribute to the detection of "dark matter"- the unseen particles not yet detected on earth, but which are believed to account for most of the mass of the universe.

8.5 SUMMARY

Apart from the various telescopes exploring different regions of the EM spectrum, there are many satellite borne observatories now orbiting different planets of the solar system and even venturing into interstellar regions in search of new discoveries. Some of these are listed below:

1. ***Juno:*** A space craft in Jupiter's orbit studying Jupiter and its various moons. Launched in August 2011, it will complete its mission by February 2018.
2. Mars and its moons has been the target of many missions using fly by, orbiter, Lander and rover space craft.
   Two Mars explorer Rovers namely Spirit, (launched in June 2003) and ***Opportunity*** (launched in July 2003), assigned to explore different regions of Mars. ***Mars Orbiter Mission called Mangalyan*** launched by Indian Space Research Organization (ISRO) is in orbit since 2014. It will explore Mars' surface features, mineralogy etc.

3. ***Cassini-Huygens:*** It is an unmanned space craft sent to orbit the planet Saturn about twenty years back. In December 2004, Huygens a rover space craft, piggy riding on Cassini landed on Saturn's moon Titan and successfully beamed data to earth using the Cassini orbiter as a relay. However, it was forced to plunge and disintegrate into Saturn's atmosphere, on commands sent from the control centre, as it had run out of fuel.
4. ***Dawn Space probe:*** launched by NASA in September 2007 with a mission for studying the dwarf planet Ceres in the asteroid belt. It is currently in orbit in spite of losing three of its four reaction wheels.
5. ***New Horizons Space craft:*** It is an interplanetary space program launched by NASA in 2006, with a primary objective to perform a flyby around Pluto in 2015 and for a secondary mission to fly further into space and study one or more Kuiper belt objects (KBO) during the next two decades.
6. ***Rosetta Mission:*** It is a space probe built by European Space Agency (ESA) and launched in March 2004 along with ***Philae*** its Lander module, assigned to make a detailed study of comet 67 P. In August 2014 the spacecraft reached the comet and performed a series of maneuvers to eventually orbit the comet at a distance of 10 Km. In November 2014, its Lander module Philae successfully landed on the comet. However, due to diminishing solar power it received Rosetta's communication with Philae was turned off in July 2016. In September 2016, Rosetta space craft ended its mission by hard landing on the comet.
7. ***Voyagers-I and II:*** The twin Voyagers-I and II spacecraft are continuing to explore regions of our solar system and beyond. Completing more than 40 year journey since their launching in 1977, they are presently, much further away from the Sun than Pluto is. Voyager-I has already made its historic entry into interstellar space. Scientists hope to learn more about this region when Voyager-II presently at the edge of the heliosphere, joins its counterpart.

To conclude, every time a new mode of measurement is employed it throws up a huge amount of new data. Analysis of this data will lead to better understanding of some of the deep mysteries of the universe.

Chapter 9

NUCLEAR FISSION VERSUS NUCLEAR FUSION

"I am become death, the destroyer of the worlds" - these famous words from the Bhagavad-Gita, as it flashed through Dr Robert Oppenheimer's mind when he witnessed an atomic explosion in July 1945"

9.1 INTRODUCTION

We are living in a world where the population is increasing at an exponential rate. It is estimated that by the year 2025, the population is likely to cross the 8 billion mark and by the end of the 21st century, it may hit the staggering figure of 12 billion. Clearly, catering to the energy needs of such a vast population, even assuming minimum living standards is a stupendous task.

Our non-renewable sources of energy such as coal, oil, and gas are likely to last only a few decades more from now on. Apart from this limitation their unbridled use will pollute the atmosphere with carbon emissions leading to global warming and consequent dangerous climatic changes. Therefore, the time has come to fully exploit the benefits of renewable sources of energy such as solar, wind and nuclear power. While we should make every attempt to fully utilize the first two sources of energy, they suffer from some severe limitations.

The first and foremost is the low energy density associated with these sources, which precludes their making a substantial impact on our vast energy requirements. The second is the intermittent nature of the energy produced, which will in turn require expensive investment in energy storage systems. The only other alternative seems to be the large scale exploitation of nuclear energy- both of the fission and fusion types. No doubt, there are disadvantages even in this case, but their benefits far outweigh the deleterious consequences associated with them.

It is in this context that we discuss in the following sections, the essentials of nuclear fission and fusion, with a comparative assessment of their merits and the position they occupy in our formulation of a global energy strategy. Before attempting this, an explanation of some of the terminologies and concepts involved seems to be in order. Such an explanation will lead to a better appreciation of the topic discussed. The explanations follow.

9.2 SOME BASIC PRELIMINARIES

To start with, there are in all 92 stable elements found in nature as listed in the periodic table. The smallest entity of an element, which enters into chemical reaction with other elements, is known as the atom. An atom is unique in the sense that it is different for different elements. The Bohr's model of the atom envisages a central nucleus composed of protons (particles carrying positive charge) and neutrons (particles carrying zero charge), with electrons (particles with negative charge) moving round the nucleus in circular orbits. Their orbital paths are however predetermined from quantum mechanical considerations. The protons and neutrons have almost the same mass, but electrons have much smaller mass which is about 1/2000 th of the mass of either the proton or neutron. In a neutral atom the number of protons in the nucleus is the same as the number of electrons orbiting it and hence a neutral atom has zero charge. The atoms are extremely small in size at about 10^{-10} meters in diameter. Except for the nucleus which occupies about 10^{-14} meters across, most of the space inside the atom is practically empty. In cases where we do not want to distinguish a proton from a neutron in the atomic nucleus, we refer to either of them by the generic term, **nucleons**.

What holds the nucleus of the atom together, when normally the positive charges of the protons in the nucleus should lead to violent repulsion as dictated by the electromagnetic force? The explanation is the presence of another force known as the ***strong nuclear force*** which is much more powerful than the repulsive force of the protons. However, the domain of this force is strictly restricted to very small distances of the order of 10^{-15} meters. Within its domain the strong nuclear force is attractive as opposed to the repulsive force experienced by the protons. Moreover, the strong force acts equally on both protons and neutrons irrespective of the charges carried by these particles. This is in contrast to the EM force which has no effect on the neutrons. One other major difference between EM force and the strong nuclear force is that the influence of the former extends right up to infinity (even though the effect becomes progressively smaller with distance), whereas the latter's influence is limited to a very small distance. It may be further noted that while the gravitational force which has a major influence on the structure of the universe has negligible effect as far as the atomic nucleus is concerned. As stated earlier, every element is distinguished by the number of protons in its nucleus. Thus the hydrogen atom has one proton and no neutrons in the nucleus, while the heaviest element namely Uranium-238(U238), has 92 protons and 146 neutrons in its nucleus. Atoms having the same number of protons but different number of neutrons are known as ***isotopes*** of the atom. For example, there are 2 isotopes for the hydrogen atom namely, Deuterium (one proton and one neutron) and Tritium (one proton and two neutrons). In this connection, the following symbols and their definitions are relevant.

1. Z (atomic number) = Number of protons in the nucleus.
2. N (neutron number) = Number of neutrons in the nucleus.
3. A (atomic weight) = $Z + N$.

The symbol, X - denotes an element X with "A" nucleons in its core out of which "Z" are protons.

The symbolism is especially useful while representing the splitting of a heavy atom into two constituent atoms (as in fission) or combining two light atoms into a single heavier atom (as in fusion). The neutron plays an important role in the structure of atomic nucleus. The strong nuclear force because of its very short range can provide attractive force only to its neighboring nucleons. On the other hand, EM force acts upon all protons in the nucleus wherever they are located, with decreasing force, when they are not in close proximity.

Neutrons present in the nucleus are influential in two ways. First, the attractive force acting inside the nucleus is exerted on both protons and neutrons. Second, their presence serves to keep the protons apart spatially, thus reducing the repulsive power of protons. The sum total of the opposite effects of nuclear force and EM force is clearly reflected in the stable structure of the nucleus. Generally in lighter elements, the number of protons and neutrons in the nucleus is almost equal, whereas for heavier elements, neutrons far outnumber protons. For example, the common isotope of carbon has 6 protons and 6 neutrons, while the heavy element Uranium-238(U238) has 92 protons and 146 neutrons. While discussing the nucleus of an atom, an important concept is the ***binding energy*** of the nucleus. In general, whenever two light elements undergo transmutation to form a heavier element, there is a loss of mass involved known as the ***mass defect*** or ***missing mass***. The missing mass corresponds to the energy involved in the transmutation process. For example, consider the deuterium atom which has one hydrogen atom (proton) and one neutron as its constituents. One should normally expect the mass of deuterium atom as calculated below to be the sum of the masses of its constituents.

But, the measured mass of the deuterium atom is only 2.014102 units, which is 0.002388 units less than the total mass. The missing mass is accounted for by the energy given off when the deuterium nucleus is formed. To digress a bit, 1 unit mass by definition corresponds to 1.66054×10^{-27} Kg, which in turn can be shown to be equivalent to 931.49 M eV. Hence the missing mass of 0.002388 units, works out to 2.2 MeV. Different elements and their isotopes are associated with different binding energies. To recapitulate, the mass of any element as listed in the periodic table is always less than the sum of the masses of protons and neutrons in its nucleus. If this deficit mass is denoted by Δm, then by Einstein's relationship ($E = mc^2$), it is equivalent to the binding energy associated with the element, ie, $E = \Delta m \times c^2$. It is the release of this energy which creates the devastating explosive force in a nuclear device. The ***binding energy per nucleon*** of an element is obtained by dividing its binding energy by the number of nucleons in its nucleus. Thus the binding energy of deuterium per nucleon is 2.2 M eV/2 namely, 1.1 M eV. Fig 1 below, gives the binding energy per nucleon as a function of atomic number A. This shows that when two light elements combine to form a heavier element, the binding energy of the heavier particle, causes energy to be given off. Alternatively, when a heavy nucleus is split into two lighter ones, the binding energy per nucleon of either of the product nuclei is greater than the binding energy per nucleon of the parent nucleus. This is pictorially explained in Fig 2.

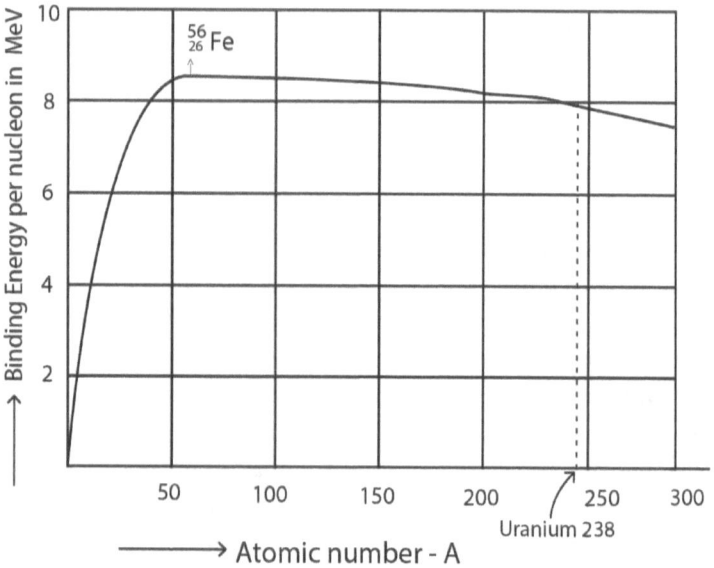

Fig 1: Atomic number versus binding energy per nucleon

Fig 1 shows how the binding energy per nucleon increases steadily till it reaches a maximum value at A = 56, which corresponds to the Iron atom. Thereafter it decreases at a slow rate up to A = 238 corresponding to natural uranium atom. This shows that when two light nuclei combine to form a heavier one (a process called fusion), the greater binding energy of the heavier one causes the energy to given off. Alternatively, when a heavy nucleus is split into two lighter ones (a process called fission) the sum of the binding energies of the product nuclei also causes energy to be given off. This is pictorially illustrated in Fig 2.

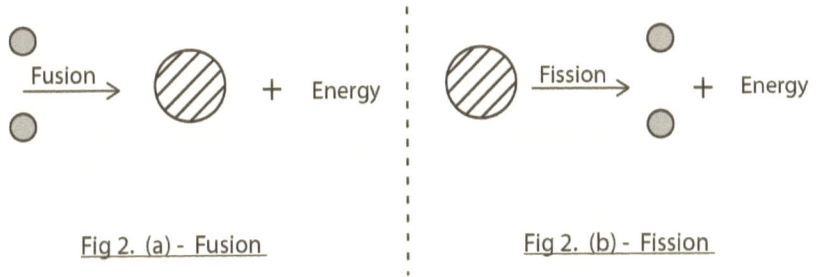

Fig 2. (a) - Fusion   Fig 2. (b) - Fission

Fig 2: Illustration of Fusion and Fission reactions

With the above preliminary concepts out of the way, we proceed to describe in more detail commercial nuclear reactors of the fission type, now successfully functioning all over the world.

9.3 NUCLEAR FISSION REACTOR

A nuclear reactor is a device used in power plants for generating electricity. Nuclear reactor converts the energy released from controlled nuclear fission into thermal energy, which is then converted to steam by heat exchangers to drive turbine generators and generate electrical power.

Fission Process: When the nucleus of an isotope of uranium namely, Uranium 235 (U235) absorbs a neutron, it undergoes nuclear fission. The resulting heavy isotope U236 is highly unstable and immediately splits into two or more lighter nuclei, releasing neutrons and kinetic energy. This energy is shared to a large extent by the neutrons released as compared to other fission products. The highly energetic neutrons bombard in turn further U235 nucleons releasing more neutrons in the process. This process is known as *chain reaction* and it goes on and on, if not controlled by external means. Fig 3 displays how the fission process is initiated in the first phase and how thereafter the chain reaction proceeds in a sequential manner.

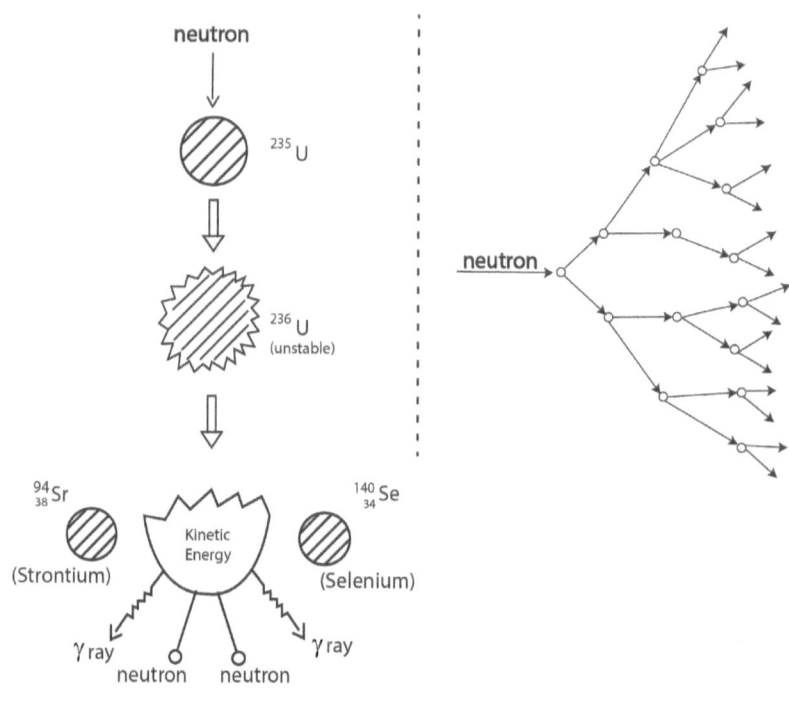

(a) Initiation of fission process   (b) Chain reaction in progress

Fig 3: Chain Reaction

Unless the chain reaction is controlled, it will ultimately result in tremendous heat production, possibly culminating in reactor "melt down". This control is achieved in two ways as explained below.

To begin with the reactor chamber consists of Uranium oxide (Uo_2) pellets filled in hundreds of long thin tubes sealed at either end, which serve as the fuel for the reactor. Control rods of Cadmium or Boron which control the nuclear reaction, can be lowered or raised into or from the reactor and are inserted side by side with the fuel rods. In a water moderated reactor, water at a very high pressure (to prevent its boiling), is circulated round the reactor core. In this fission process, the water acts both as a moderator and coolant. It serves in slowing down the fast neutrons resulting from the fission process apart from carrying away the heat generated to a heat exchanger, to produce steam which in turn drives the turbine-generator. In most cases the cooling system is physically separated from the water that is used to raise steam from the point of view of safety. The rate of fission reaction is primarily controlled by the control rods which are remotely moved up or down in the reactor vessel. When the control rods are inserted deep inside the core, it absorbs neutrons. Hence there are fewer neutrons to cause chain reaction and consequently the power output decreases. Alternately, when the rods are lifted up, the rate of fission increases, resulting in increased power output.

A water moderated light water reactor using enriched uranium fuel may contain about 90 tonnes of Uo_2 to yield 1000 MW of electric power and the fuel needs to be replaced only every few years as and when the U-235 content is used up. The most commonly used reactors are Light Water Reactors (LWR), which uses 3 to 5 per cent U235 enriched fuel. These reactors use ordinary water as coolant and moderator.

The Pressurized Heavy Water Reactor (PHWR) which is not so common these days uses slightly enriched Uranium which may contain about 1.1 per cent of U235. These reactors use Heavy Water both as coolant and moderator. The naturally abundant fuel found in nature namely U238, contains only 0.7 per cent of U235. The enrichment of fuel is done in special plants using high-speed gas centrifuges.

9.4 NUCLEAR FUSION REACTOR

Nuclear fusion reaction is what happens in the sun and the stars and the mechanism responsible for all chemical elements present on earth. It is the process in which two light nuclei bind together to form a heavier nucleus. Since the mass of the heavier nucleus is slightly less than the total mass of the initial nuclei forming it, the mass deficit is released as energy.

In the sun, energy is released through the fusion of hydrogen atoms to form helium. It occurs in stages with the fusion of two hydrogen atoms (comprising 2 protons) into a deuterium atom (one proton and one neutron). The deuterium atom combines with another hydrogen atom to form helium-3 nucleus (two protons plus one neutron). In the next stage the helium-3 nucleus fuses with another helium-3 to form helium-4 (two protons plus 2 neutrons) accompanied by release of energy. In the sun, this process of energy generation lasts hundreds of millions of years. It is fortuitous that it is so because otherwise the hydrogen fuel in the sun would burn out in a few million years, hardly giving a chance for life on earth to evolve.

However, there is a quicker rate at which fusion can occur, involving deuterium (De-2) and tritium (Tr-3), both isotopes of hydrogen. Because this arrangement involves only a rearrangement of protons and neutrons and not the transformation of a proton into a neutron (as in the case of the sun), the fusion reaction proceeds much more rapidly. The price we pay for this modification is that less energy is released than in the case of the sun. One other problem here is that the deuterium and tritium will fuse together only at very high temperatures of 100 million degrees Kelvin which is 10 times more than the temperature existing in the nuclear fusion taking place in the sun. To achieve this, the reactants are initially heated so that their combination is split into positively charged ions and electrons. This hot plasma (ionized gas) thus produced, should be confined long enough at a very high density and temperature, for fusion to occur. In the sun, the confinement of the plasma is effected by the enormous inward pull of the gravitational field of the sun. On earth the repulsive force of positive ions have to be contained to make them come closer together. Further, there must be a container to contain this hot plasma and any physical container we could think of will melt at such high temperatures. One of the proposed methods for confining the hot plasma without the plasma coming into contact with the walls of the container is by what is known as "magnetic confinement". In this method, magnetic fields hold the charged plasma particles together at the center of a highly evacuated and sealed container vessel known as a toroidal chamber. By far the most promising of the various containment methods suggested, is via the concept of "Tokamak" proposed by Soviet physicists. The Tokamak is a doughnut shaped vessel or torus, in which a helically spiraled magnetic field is artificially created which insulates the charged particles of the plasma from the surrounding walls. The twisted helical field inside the torus is generated by the interaction of two magnetic fields namely, the toroidal field and the poloidal field as shown in Fig 4.

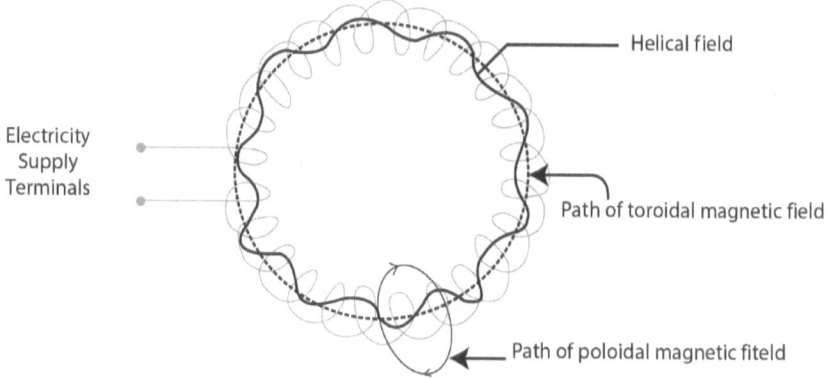

Fig 4: Structure of Tokamak
The Toroidal and Poloidal magnetic field in combination produce a helical magnetic field confining the Plasma.

The toroidal field which acts the long way around the torus is produced by coils wound round along the torus as in a solenoid which generates a pulsating magnetic field. The plasma inside the torus, being a conducting medium, acts as the short circuited secondary of a transformer due to magnetic induction. The resulting current flowing through the plasma serves two purposes - one it heats the plasma and the other it produces a poloidal magnetic field, which acts at right angles to the existing toroidal field. The interaction of these two fields produces the desired helical confinement field. The hot plasma whose tendency is to escape outward is thus trapped inside the magnetic field. In the graphic words of an author "it tends to twist and turn like an enraged snake struggling to escape".

The Tokamak concept is sought to be implemented on a large scale, in a fusion reactor model being built at Cadarache in France, supported by an international team of scientists and technologists from a number of countries like European Union, China, Russia, Japan, South Korea and India. Known as ITER (*International Thermonuclear Experimental Reactor*) the project which was conceived in 2006 has suffered several cost and time overruns. It is now expected to cost US $50 billion (about 10 times the original estimate) and the pilot plant of about 400 MW of electricity, is supposed to be ready by 2027, almost 11 years behind the earlier schedule. The main features of ITER are as follows.

The ITER uses deuterium and tritium as fuel. The reaction involves the fusion of a deuterium and tritium nuclei to form a helium nucleus (an alpha particle) plus a high energy neutron as indicated below:

Deuterium + Tritium ⟶ Helium 4 (α particle) + Neutron + 17.6 MeV
^{2}D + ^{3}T ⟶ ^{4}He + 1 neutron + 17.6 MeV

From Fig 1, it is evident that all stable isotopes lighter than Iron-56 (which has the highest binding energy) per nucleon, will fuse together and release energy. However in fusion deuterium and tritium are found to be the most suitable for energy generation, as they require the lowest activation energy (thus the lowest temperature) to do so, while at the same time producing the most energy per unit.

The optimal fusion temperature required for Deuterium-Tritium fusion reaction to take place is about 100 million degrees Kelvin. The expected electrical output for ITER is about 400 MW. As explained earlier the nuclear fuel is held inside a toroidal ring shaped reactor called the tokamak. Super conducting coils of niobium alloy weighing a total of 10,000 tonnes and cooled by liquid helium carry current which produces the toroidal field. The fusion fuel is simultaneously heated in three different ways. While the electrical circuits force a current through the plasma, it is also blasted with microwaves and bombarded by high energy atoms generated by small particle accelerators, dotted round the toroidal ring. A lot of power has to be pumped in to make the plasma sizzling hot to start with, expecting that 10 times more power will emerge as the output. The fusion reaction will create high-energy neutrons, which being electrically neutral cannot be contained by the magnetic cage. These neutrons will escape the cage and bombard the walls of the containment vessel which is made up of half a meter thick stainless steel blocks, with high pressure water pipes embedded inside. The water flowing in these pipes is designed to carry the heat away to the heat exchanger. The inner surface of the steel container is not strong enough to withstand the bombardment of neutrons, which will chip off iron atoms of the steel container. These will in turn contaminate the fuel and dampen the fusion reaction. To minimize this damage, the inner walls of the container are lined with beryllium tiles. Apart from neutrons, the main by-product of the fusion reaction is helium nuclei, which also stores a part of the energy produced by the fusion process.

This will continue to remain inside the magnetic cage and contribute to further heating of the plasma. However, ultimately they have to be removed as helium ash accumulation will contaminate the plasma and slow down the fusion reaction. This is achieved by means of a "***divertor***" located at the bottom of the reaction chamber. The divertors job is to skim the outermost layer of the plasma and remove the helium "ash" and other impurities. The divertor is covered with tungsten and carbon fiber whose melting point is about 3000 degrees Kelvin. If the ITER is able to successfully complete the fusion project as proposed, it will provide humanity with a potentially safe and environmentally friendly energy source with almost unlimited capacity.

9.5 FISSION VERSUS FUSION – A COMPARISON

In a world virtually starved of non renewable sources of energy, in the not too distant future, the nuclear power generated by fission and fusion will come as a big boon.

Fission: There are at present more than 400 and odd atomic power stations working in different countries generating about 200,000 MW of electric power. This is equivalent to more than 10 billion barrels of oil per day. Countries like France and Belgium obtain more than half their electricity from atomic power even now. However, three catastrophes associated with fission reactors have considerably dampened the enthusiasm of governments to install new atomic reactors. First in 1979 March, The Three Mile Island nuclear station disaster in Pennsylvania, USA destroyed one of the reactors resulting in certain amount of radioactive material escaping into the atmosphere. The second relates to the more serious Chernobyl nuclear station reactor melt down which occurred in April 1986, destroying a 1000 MW reactor in Chernobyl, Ukraine. This disaster led to large loss of life and contamination of the environment due to escape of deadly radioactive matter into the atmosphere and into the ground. In the succeeding months and years following the tragedy many thousands more became ill as a result of prolonged exposure to the harmful radiation. This also continued to contaminate the food and water supply in the area. More recently in March 2011 in Fukushima, Japan the Fukushima nuclear power station was devastated by a giant tidal wave following an earthquake, resulting in reactor melt down which caused highly radioactive material to escape into the environment and contaminate the surroundings. Six years after the disaster, the neighboring land is still contaminated by the radioactive fall out and rendered uncultivable and unlivable. It is estimated that the entire clean up will take over forty years involving billions of dollars. It is unfortunate that these disasters happened, but more stringent operating procedures and safeguards recommended since then, are likely to gradually restore confidence among people and governments to set up more nuclear power stations. Another worrisome aspect of nuclear power generation is the proper and safe disposal of radioactive spent fuel from reactors. Even if a certain percentage of the nuclear waste can be reduced using reprocessing techniques, the problem of nuclear waste disposal still remains. It is relevant to point out here that the reprocessing of spent fuel yields highly radioactive plutonium which can be misused by belligerent nations to produce atomic and hydrogen bombs. Unless regulating mechanisms are in place, this poses a grave danger to humanity.

Fusion: As opposed to fission, a fusion reactor has several advantages. A few of them are listed below.

1. Tokamak requires enormously high temperature with high density plasma to sustain a reaction. The moment these conditions are diluted, fusion comes to an abrupt end. It is precisely this absence of "runaway" condition that makes fusion reactor a much more attractive proposition than its counterpart.
2. Fission reactors depend on the supply of Uranium fuel which countries have to import to keep the reactors functioning. This is a serious limitation. However, in fusion reactors, the fuels used are deuterium and tritium. The supply of the former is almost unlimited and can be extracted from water via electrolysis. The latter can be bred within the fusion reactor itself.
3. In fusion reactors there is hardly any radioactive waste to be disposed off as in the case of fission reactors.
4. As no fissionable by-products are produced, the danger of using them for making nuclear weapons is almost, nil.

9.6 CONCLUSION

If ITER succeeds in designing and operating a viable fusion reactor which can supply cheap and dependable energy, then fusion is the choice for the future. However, such an occurrence appears to be a long way off and nothing can be predicted at present with any certainty.

Chapter 10

RELATIVISTIC THEORY

"Newton and Newton's laws lay hid in the night.
God said, let Newton be and all was light"

– Alexander Pope

10.1 INTRODUCTION

In the year 1905, twenty six year old Albert Einstein, then working as a junior expert at the Swiss patent office in Berne published three path breaking papers. The first paper on photo electric effect, proposed that light has a dual character possessing both particle and wave properties. The subject of the second paper was Brownian motion- the irregular zigzag movement of tiny bits of suspended matter such as pollen grains in water. Einstein showed that Brownian motion results from the bombardment of particles by randomly moving molecules in the fluid in which the particles are suspended. In the third paper he introduced the special theory of relativity. He followed it up in 1916 by his work on the general theory of relativity, relating gravity to the structure of space time. Paradoxically, Einstein was awarded the 1921 Nobel Prize in physics, not for his work on relativity which he considered as most important, but for his discovery of the law of photoelectric effect

10.2 NEWTONIAN MECHANICS

Before the advent of Einstein, Isaac Newton was virtually worshipped as a God in the world of physics. His unified law of gravity was simple and equally applicable to small and large masses of matter. His inverse square law explained in an elegant mathematical manner, the elliptical orbits of the planets around the sun and Kepler's laws governing their motion. One of the stunning successes of Newton's theory related to the orbital motion of planet Uranus. Earlier, astronomers were at a loss to explain the strange perturbations noted inthe orbital path of planet Uranus. However, calculations based on Newton's gravitational law predicted the presence of an unseen planet which was the main cause of the unexplained perturbations. This planet was subsequently identified as planet Neptune. Thus the existence of of a planet was predicted much before it was actually observed using Newtonian mechanics. However, astronomers were faced with another challenging problem soon after, regarding the precession of the perihelion of planet Mercury. Normally a planet will go round and round endlessly in an elliptical orbit round the sun. In the case of Mercury, it was noted that its perihelion shifted by approximately 4 arc seconds every century (an arc second is 1/3600 th of a degree), due to unknown causes. Newton's laws were unable to explain this anomaly. However, when Einstein proposed his theory of relativity, one of the test cases for his theory was the explanation of Mercury's anomalous orbit. By applying the general theory of relativity, Einstein was able to conclusively show that the observed value of the precession of Mercury was in entire agreement with the value predicted by his theory. Subsequent events showed that the universal law of gravitation as proposed by Newton was only an approximation and inadequate while dealing with the very high gravitational fields. This provided the motivation for Einstein to propose his theory of relativity.

10.3 SPECIAL THEORY OF RELATIVITY

To start with one has to be clear about what one means by motion. A passenger moves relative to an airplane and the plane moves relative to earth. The earth in turn moves around the sun and the sun moves in our galaxy and so on. In each case a frame of reference is important. In this connection, an *inertial frame of reference* is defined as one that moves with a constant velocity (constant both in speed and direction). All reference frames that moves at different constant velocities relative to an inertial frame of reference are also inertial reference frames. Consider a train moving at constant velocity and a person sitting inside the train with all its windows closed. There is no way for the person to find out whether he is moving or stationary. This is because, if you conduct any experiment in one inertial reference frame, you will get the same result in any other inertial reference frame. Thus laws of physics are identical regardless of the inertial reference frame. The special theory of relativity as proposed by Einstein in 1905 deals only with inertial frames of reference. Later on in 1916, Einstein proposed a more comprehensive theory which dealt with accelerated reference frames.

The special theory of relativity rests on two fundamental postulates. They are:

1. The laws of physics must be the same in any inertial reference frame.
2. The speed of light has always the same value of 3×10^8 m/sec, in any inertial frame.

The first postulate is justified by countless experiments performed to test its validity. The second postulate needs some explanation. In Newtonian universe, speed is additive. For example, in Newton's world, if one compares the speed of light emanating from a stationary light source located on a railway platform, to a light emanating from the head light of a speeding train passing by the platform, the speed of light appears faster and therefore exceeds its speed in free space. But in Einstein's universe, no vehicle or device fitted with a light source, however fast it may move can ever break the speed of the light in free space. Two results flow from the second postulate. These relate to **time dilation** and **length contraction**. These are explained in the next section.

10.4 TIME DILATION AND LENGTH CONTRACTION

To dilate means to become larger and by time dilation we mean stretching of time. The two phenomenons arising out of the special theory of relativity are dilation of time and contraction of length.

Time Dilation: Measurements of time intervals are affected by relative motion between an observer and what is observed. Time measurement is normally done by a clock which may be any device subject to uniform repetitive cycles of change. Accordingly, time can be measured by the number of cycles, between events. The general definition of a clock implies that it may be an electrical or atomic clock or even a biological clock, where time interval between heart beats may be considered as an event. The time viewed in a reference frame where the clock is at rest is called the *proper time* of the clock. "Proper" does not mean anything special about the reference frame, but merely indicates the reference frame in which the clock is at rest. The effect which states that a moving clock ticks more slowly than a clock at rest, as viewed from the rest frame is known as *time dilation*.

To visualize how this dilation occurs, consider two identical clocks (in this case, represented by mirrors) as shown in Fig 1. In each clock, a pulse of light is reflected back and forth between two parallel mirrors distant "Lo "apart vertically. Whenever light from the lower mirror gets reflected by the upper mirror and then returns to the lower mirror, let us assume that a marker or recorder attached to the mirror records it as an event on tape.

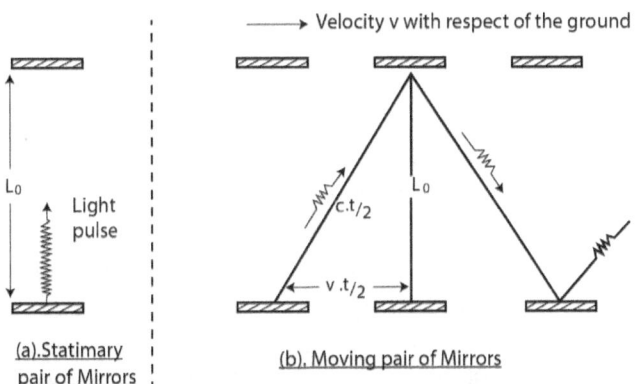

Fig 1: Stationary and Moving clocks

Assume that an observer in the ground laboratory watches both the clocks- one at rest in the laboratory and the other located in a space craft traveling at a uniform velocity "v", parallel to the ground.

In the case of the stationary mirror shown in Fig 1(a), time taken for the pulse to travel up and down is given by $t_o = 2 L_o/c$, where c is the velocity of light.

In the case of the moving mirror as shown in Fig 1(b), which moves at an uniform velocity "v" with respect to the ground, let the time interval observed from the ground as a result of reflection from the moving mirror be "t". Note that because the clock is moving, the light pulse as seen from the ground follows a diagonal path. The distance of the path of the light pulse from the bottom mirror to the top mirror is therefore c.t/2, where "c" is the velocity of light. Meanwhile the mirror pair has traveled a distance v (t/2), where "v" is the velocity of the traveling mirror pair. As is clear from Fig 1(b), we have

$$c^2(t/2)^2 = L_0^2 + v^2(t/2)^2$$

From which we obtain

$$t = \frac{2\,L_0/c}{\sqrt{1 - \frac{v^2}{c^2}}} \quad \text{since } 2\frac{L_0}{c} = t_0$$

We have

$$t = \frac{t_0}{\sqrt{1 - \frac{v^2}{c^2}}}$$

Now define

$$\gamma = \frac{1}{\sqrt{1 - \frac{v^2}{c^2}}}$$

Hence

$$t = \gamma\, t_0$$

Time dilation is not a mere theoretical concept, it can be observed in the laboratory. Consider for example, the case of a fundamental particle called muon which is 207 times the mass of the electron. They are created in the upper atmosphere of earth, when high energy cosmic rays collide with atmospheric nuclei. Muon particles are highly unstable when stationary and have an average life of 2.0 micro seconds before it is transformed into an electron. If for example the muon particle is accelerated to a speed of say 0.995c in a particle accelerator, its average life time as observed in the laboratory jumps to 20 microseconds – a dramatic increase by a factor of 10 in the case of an observer attached to the muon. The average life time is still 2 microseconds, but for the laboratory observer it is 20 microseconds. In a thought experiment, if a human being were to travel at the same speed as the muon, his life expectancy will be increased by a factor of 10 as seen from the earth. Thus, if his life expectancy is 70 years on earth, it will be increased to 700 years in the above example. This is the concept behind time dilation.

Length contraction: Relative motion affects not only time but length as well. Known as the Lorentz contraction, the length "L" of an object in motion as measured by an observer in the laboratory will always be less than its proper length "Lo". Where, "Lo" is the length of the object in the rest frame. As in the case of time dilation, it can be shown that,

$$L = L_0 \sqrt{1 - v^2/c^2} = L_0/\gamma$$

Since "γ" by definition, is always greater than unity, "L_0/γ" will be less "L_0" and hence the contraction predicted. It may be noted that if the object moves at right angles to the direction of velocity "v", then there is no contraction and $L = L_0$.

10.5 MASS ENERGY EQUIVALENCE

It is not merely length and time that are affected by theory of relativity. If "m" is the relativistic mass and "m_0" the rest mass, it can be shown that $m = \gamma m_0$.

As can be seen $\gamma = 1$, when velocity of the mass namely, "v" is zero and hence $m = m_0$ which is what we normally assume when an object travels at low velocities ($v \ll c$).

In a similar fashion, the relativistic momentum "p" is given by

p = (relativistic mass x velocity)

= $(\gamma m_0) v$

These relationships lead us to the celebrated and oft quoted Einstein's equation $E_0 = m_0 c^2$ where "E_0" is the rest energy, "m_0" is the rest mass and "c" the velocity of light.

When the body is in motion, the total relativistic energy "E" is given by

E = Rest energy + Kinetic energy.

Since $E = (\gamma m_0) c^2$ we have,

$(\gamma m_0) c^2 = m_0 c^2$ + Kinetic energy.

Hence kinetic energy = $(\gamma - 1) m_0 c^2$

The mass-energy equivalence namely, $E_0 = m_0 c^2$ plays a key role in many problems in physics. For example in a closed physical system with no external forces acting, conservation of energy is preserved in the sense that the energy considered is the total energy of the rest mass plus kinetic energy inside the system.

Another important result is the so called ***energy momentum relationship***. In a relativistic situation when "v" is very large and consequently "γ" is much greater than unity. The relativistic energy $E = (\gamma m_0) c^2$ and the relativistic momentum given by $p = (\gamma m_0) v$.

From these two relationships after substituting the value of

$$\gamma = \frac{1}{\sqrt{1 - v^2/c^2}},$$

It can be shown that $E^2 - p^2 c^2 = (m_0 c^2)^2$

Hence, $E^2 = (m_0 c^2)^2 + p^2 c^2$, i.e. (Total Energy)^2 = (Rest Energy)^2 + (Momentum)^2

The question now is whether a particle which has zero mass i.e. $m_0 = 0$ can still exhibit properties like energy and momentum. In classical mechanics the rest mass must be non-zero to have energy and momentum. But as the above equation suggests, it is possible for "m_0" to be zero (as in the case of photon) and still have energy $E = pc$, provided the particle travels with a velocity "c". This is precisely what happens in the case of a photon which has zero mass and travels at a velocity "c" and carries energy packets depending upon the frequency of radiation.

10.6 GENERAL RELATIVITY

Special relativity is only concerned with inertial frames of reference. But Einstein's 1916 general theory of relativity goes one step further by including the effect of accelerated reference frames. Our understanding of gravity is encapsulated in Einstein general theory of relativity. It supposes that space and time together form a dynamical manifold whose curvature influences the motion of matter according to ***the principle of equivalence***. This principle is central to general relativity. According to this principle an observer in a closed laboratory cannot distinguish between the effects produced by a gravitational field in the laboratory and that produced by the acceleration of the laboratory itself. For example, assume a ball dropped from a height "h" inside a stationary laboratory, (with only earth's gravity "g" acting on it), hitting the ground in a time "t" sec. The same laboratory if transported to outer space with zero gravitational pull acting on it, but which is traveling with an acceleration "g" imparted to it will record exactly the same time "t" sec if the experiment is repeated there. This experiment in a way demonstrates the principle of equivalence. We can further conclude from this experiment that *the gravitational mass of a body is identical to its inertial mass*. According to the general theory of relativity, under the influence of a very high gravitational field even light which normally travels in a straight line suffers a change in direction. For example, light rays from stars which graze the sun's surface gets bent towards it by an angle of about 0.005 degrees according to general theory of relativity. The veracity of this assertion was strikingly demonstrated by an experiment conducted by astronomer Arthur Eddington and his team in 1919, when they measured the bending of light from the stars, during a total solar eclipse. This angle of deviation was found to be in close agreement with what the theory predicted.

Another phenomenon relates to what is known as ***Gravitational Lensing***. In this case a bright light source for example a quasar even though it is hidden from earth's view by a massive body such as a galaxy, becomes visible to us because of the bending of light, effected by the massive body. This bending of light is analogous to parallel beams of light incident on a convex lens, which in turn converges the beams to a focus in front of it.

Another discovery of great significance predicted by Einstein about 100 years ago, relates to what is known as ***Gravitational Waves***. These are ripples in the fabric of space-time generated by violent events such as collision of massive objects like black holes or galaxies. The waves travel across the universe at the speed of light alternately stretching and squeezing space-time as they go. The amplitude of these waves when they reach the earth is exceedingly small- tens of millions of times smaller than the size of a single atom. Hence the great difficulty in detecting these waves. However, in recent times the development of sophisticated instruments such as the LIGO (Laser Interferometer Gravitational wave observatory) and VIRGO detectors (these are "L" shaped detectors which track gravitational waves using the principle of laser interferometry), have enabled scientists to detect these waves. The most recent observations recorded the merging of two massive black holes each about 30 times the mass of the sun and located 1.3 billion light years away from us. This discovery will have far reaching consequences and will open up future opportunities to learn more about black holes, neutron stars, and other astronomical bodies. It will also enable us to look deeper into the universe, almost a few micro seconds after the big bang. The detection of gravitational waves was considered to be so important that the 2017 Nobel Prize in physics was jointly awarded to three scientists for their pioneering work in this area. In a way, this award may be considered as a celebration of the genius of Albert Einstein, who first predicted the existence of gravitational waves.

To conclude, every observable phenomenon predicted by the general theory of relativity has been meticulously verified. Indeed, the special and general theory of relativity constitutes the major underpinnings of modern day physics.

Chapter 11

QUANTUM MECHANICS – A PARADIGM SHIFT

"The great tragedy of science- The slaying of a beautiful hypothesis by an ugly fact"
— Thomas Henry Huxley

11.1 INTRODUCTION

Twentieth century witnessed the birth of two of the most revolutionary concepts in physics, namely "Theory of Relativity" and "Quantum Mechanics". While the former was entirely the brain child of Albert Einstein, quantum mechanics was developed and fostered through the joint efforts of a host of brilliant physicists like Max Born, Werner Heisenberg, Paul Dirac and Erwin Schrödinger.

The theory of relativity worked extremely well when applied to macroscopic phenomenon. However, it failed to yield satisfactory results when applied to the subatomic world of atoms and elementary particles. A truly different approach was called for and quantum mechanics fitted the gap in admirable manner. At first sight, quantum theory or quantum mechanics as it is popularly known appears to be weird and abstract even to some of its most ardent votaries, so much so that the renowned physicist Richard Feynman was supposed to have once remarked "Nobody really understands quantum theory". Yet over the years from 1925 onwards, it has proved to be one of the best tested theories ever known, with remarkable agreement between theoretical predictions and observational facts. It has since proved its usefulness in many applied areas such as micro electronics, computers and medical diagnostics.

11.2 CLASSICAL PHYSICS VERSUS QUANTUM MECHANICS

Physicists refer to an object that absorbs all its infalling radiation, regardless of its energy content as a *black body*. If the black body maintains a constant temperature, the energy it emits is known as *black body radiation*. Till the German physicist Max Planck came on the scene in 1900, classical physics failed to predict the way in which the intensity of black body radiation varied with frequency. Classical theory actually predicted that at higher frequencies of incident radiation the intensity of black body radiation goes on increasing. However, experiments showed the converse to be true with the intensity of black body radiation actually decreasing at higher frequencies of incident radiation. To correct this anomaly, Planck hypothesized that energy came in quanta or packets, the size of the packet depending only on the frequency involved.

This is given by the simple relationship,

$$E = n \, (hf),$$

where "E" is the energy, "h" the so called Planck's constant, "f" the frequency and "n" takes on integral values. Incorporating this idea, Planck's radiation formula exactly matched observational data.

The next major development in quantum mechanics took place in 1905 when Albert Einstein published his seminal paper on photoelectric effect. Prior to the publication of this paper, scientists were already aware that when light of high intensity impinged on the surface of certain metals, electrons were liberated. But the underlying theory of why this happened and that too at certain frequency of the light photons baffled understanding. In order to solve the problem, Einstein hypothesized that light instead of being considered as a wave can also be thought of as made up of particles called photons each carrying packets of energy (hf) depending upon the frequency associated with it. He even went a step further and associated this phenomenon (known as ***photoelectric effect***), to the work function of the metal involved. Incidentally, by way of explanation, ***work function (W)*** is the amount of energy required to liberate an electron from the atomic forces holding it back. When light impinges on the metal surface, Einstein linked the kinetic energy (KE) of the liberated electrons with the work function (W) by the relationship,

$$K E = (hf) - W$$

In order to liberate electrons at a particular frequency, the quantum energy (hf) has to be always greater than the work function (W). No amount of light intensity impinging on the metallic surface however high it may be can liberate electrons unless it is delivered at a frequency higher than the threshold frequency suggested by the above relationship. Once this frequency is crossed, increasing numbers of electrons can be liberated by increase in light intensity. For this remarkable piece of work Einstein received the Nobel Prize in physics in 1921. The two examples sited above show conclusively the power and versatility of the quantum mechanics approach. When applied to physical problems, this approach demonstrated for the first time the concept of ***wave-particle duality***. More examples followed in subsequent years.

11.3 DE BROGLIE WAVES

In 1924, Louis de Broglie a physicist from France argued that there is a symmetry in the wave-particle duality relationship. If waves could be treated as particles on occasions, why not particles exhibit wave properties?. He hypothesized that there exists a relationship between wave property associated with "λ" the wave length and particle property associated with its momentum p = mv namely,

$$\lambda = h/p$$

The wave length "λ" as defined above, is known as the *de Broglie wave length*.

According to this concept, *every moving object* is associated with its own wave length depending upon its mass and velocity. Thus a spherical ball of mass 1 Kg moving with a velocity 1 m/sec, has a value

$$\lambda = h/mv = 6.6 \times 10^{-34}/(1) \times (1) = 6.6 \times 10^{-34} \text{ m}$$

where, h = 6.6×10^{-34} is the Planck's constant.

As can be seen, the value of "λ" in the above case is infinitesimally small compared with the size of the object and hence its effect on the object is almost negligible. But this is not the case when applied to an electron with a mass of 9.8×10^{-31} Kg, moving with a velocity of say 10^6 m/sec. Note that the speed chosen in this case is considerably less than the velocity of light and hence we can safely assume that the rest mass of the electron is the same as its relativistic mass. Hence we have,

$$\lambda = h/p = (6.6 \times 10^{-34})/(9.8 \times 10^{-31}) \times 10^6 = 7 \times 10^{-10} \text{ m}$$

Note that in this example, the wave length is comparable to the size of the electron and one is justified in treating electron motion as a wave motion and not as a particle.

11.4 THE BOHR MODEL OF THE ATOM

The British physicist J.J Thomson, who discovered the electron in 1898, envisaged the atom as a fruit cake with positive charges uniformly distributed in matter and electrons embedded in it like resins in the fruit cake. This model was overturned by Ernest Rutherford a New Zealand born physicist working at Cambridge, when he proposed an atomic model analogous to that of the solar system with the nucleus of the atom composed of protons as stationary, just like the sun and the electrons like planets revolving round. While this was certainly a step forward from that proposed by J.J Thomson, its validity was open to serious question as the conclusions arising from the model were in conflict with the existing principles of physics. Specifically, the electrons while circling the nucleus (because of its continuous change of velocity), are subject to accelerated motion and therefore continuously emitted energy in the form of electromagnetic waves. Because of this, in a brief time, electrons will lose energy and go into a spiraling motion and crash into the nucleus leading to collapse of the atom. As this was not happening, it was soon clear that something was amiss in the Rutherford model. It was at this juncture that Neils Bohr a physicist from Denmark took up the challenge. Bohr soon realized that it was hopeless to tackle the problem from the point of view of classical physics. In 1913, he altered the picture by incorporating ideas from quantum mechanics. He argued that electrons cannot orbit the nucleus in any arbitrary fashion and that quantum mechanical considerations demand that only certain specific orbits are permissible. He also showed that when electrons moved along these orbits, there would be no radiation and stability is preserved. Each of these permissible orbits are assigned numbers known as **quantum numbers**. Bohr's model can be used to determine the radius of the innermost orbit of the hydrogen atom (made up of one proton in the nucleus and one electron orbiting it). This radius is approximately 0.529×10^{-11} m. Almost a decade after Bohr's model was proposed, Louis de Broglie proposed an alternate way of arriving at Bohr's results using the concept of de Broglie waves. The results turned out to be exactly similar to that of the Bohr's model.

We recall that the de Broglie wave length is given by $\lambda = h/mv$. In terms of the de Broglie wave length, it is possible to make the following statement for the hydrogen atom.

"An electron can circle the nucleus of an atom only if the circumference of the orbit is an integral multiple of de Broglie wave length"

Specifically, if λ is the de Broglie wave length defined by "h/mv", the condition for orbital stability states,

$n\lambda = 2\pi r_n$ where, n=1,2,3,.........etc, represents the respective electron orbits

Thus the second permissible orbit of the electron should have a circumference exactly 2 times λ and so on. The corresponding radii for different orbits can be obtained, from the formula,

$r_n = n^2 r_1$ where, $r_1 = 5.3 \times 10^{-11}$ m and n= 1,2,3......etc,

Each of the permissible orbits is associated with respective energy levels E_1, E_2, E_3....... etc. The lowest energy level E_1 is associated with the nearest orbit circling the nucleus and has a value of -13.6 eV. This corresponds to n=1. The general rule for energy levels is given by,

$$E_n = E_1/n^2$$

where, E_n is the nth energy level, E_1 the first energy level and "n" the corresponding integer.

Thus $E_2 = (-13.6)/2^2 = -3.4$ eV. Thus, $E\infty$ corresponds to zero electron volts. To start with the electron of the hydrogen atom in the first orbit is bound to the atom with a certain force acting on it and the energy expended to release it from this bond to attain higher levels of energy, is shown in Fig 1 of chapter 4. The figure also shows graphically, the various orbital levels and energy associated with these levels.

Note that the energy levels are all negative which indicates that the higher the energy level, the less negative it will be. An *atomic electron* is stable only at those energies. The lowest level E_1 is called the **ground state** and the higher levels E_2, E_3 etc, are called the **excited states**. At E_∞ the electron is no longer bound to the nucleus and is completely free. The work required to remove an electron from the ground state to its free state is known as **ionization energy**, which is 13.6 eV for the hydrogen atom. When an electron jumps from a lower state E_i to a higher state E_j due to external stimuli, there is a quantum jump of energy namely, $(E_j - E_i)$ whose magnitude according to quantum theory exactly equals (hf). That is,

$$|E_j - E_i| = (hf)$$

The frequency of radiation involved is governed by the equation,

$$f = |E_j - E_i|/h$$

The similar thing happens when the electron returns to its lower energy state from a higher energy state, with the electron emitting radiation. The alternate explanation of the Bohr's model for a hydrogen atom using de Broglie waves is a clear example of how wave theory can come to the rescue of classical physics, when the latter is unable to explain certain phenomenon.

11.5 HEISENBERG'S UNCERTAINTY PRINCIPLE

In 1927, the German physicist Werner Heisenberg postulated the **principle of uncertainty** that bears his name. It is one of the most important ideas in quantum mechanics. It states that,

$$\Delta x \, \Delta p \geq h/2\pi$$

where, Δx indicates uncertainty in position and Δp uncertainty in momentum and "h" the Planck's constant.

According to this relationship, the product of these two uncertainties should exceed $1/2\pi$ times the Planck's constant. In other words, the more accurately we try to fix the position of an object, the less accurate we will be regarding its momentum. This is just not a matter of having inadequate equipment for the purpose of measurement but the statement goes much deeper. In fact it has direct links with the wave-particle duality which we discussed earlier. Consider for example, the attempt to observe an electron using visible light. Compared to the size of an electron, the wave length of visible light is so huge that we will completely miss locating the electron. Indeed we need an extremely short wave length of radiation to even locate it. But short wave length implies very high frequency, as indicated by the relationship between wave length and frequency ($c = \lambda f$ or $f = c/\lambda$). Clearly, high frequency implies high energy as given by the relationship $E = hf$. Such high energy impinging on an electron will seriously affect its position and velocity. If on the other hand we use a low energy photon with a large wave length to probe the electron, we will not even be able to locate its position accurately. One important aspect of measurement is that in order to observe something we must interact with the object we are trying to observe. Whatever means we use for the observation, the very act of observing the object will disturb it and vitiate the accuracy of measurement. The interpretation of the uncertainty principle can be extended to other parameters also, like time and energy.

Thus we have,

$$\Delta t\, \Delta E \geq h/2\pi,$$

which states that if energy is measured in an arbitrary time interval which is very small(denoted by Δt), such energy measurement will turn out to be extremely inaccurate.

Closely related to the principle of uncertainty is the exclusion principle proposed by Wofgang Pauli in 1925. Broadly speaking, this fundamental principle states that,

No two electrons of an atom can exist in the same quantum state.

The central idea here is that two identical spin (1/2) particles, within the limits of the uncertainty principle cannot both have the same position and the same velocity

11.6 THE WAVE FUNCTION AND SCHRÖDINGER'S EQUATION

The success of Newton's laws and other physical principles during the nineteenth century, led to a firm belief in scientific determinism among practicing scientists. In classical mechanics if we know the position and velocity of a particle along with the forces acting on it, at a particular time we can precisely predict its future trajectory at all times. This even led the French mathematician Marquis de Laplace to assert that if we know the position and velocities of all particles in the universe at any particular time, the laws of physics should enable us to calculate the state of the universe at all times. But this sweeping assertion was overturned by Heisenberg and others when they proposed the uncertainty principle. When the accuracy of the initial conditions themselves are in doubt, how can one predict the subsequent evolution with certainty, they argued. This is especially true while dealing with particles at an atomic level.

In quantum mechanics a particle does not have a well defined position or velocity (unlike its counterpart in classical mechanics), and the state of the system is represented by what is known as the ***wave function***. The wave function assigns a number at each point in space, which gives the probability that the particle can be found in that position. Some wave functions are sharply peaked at a particular point in space. In such cases there is only a small uncertainty in the position of the particle. But here the wave function changes rapidly in the neighborhood of that point. This means the probability distribution of velocity is spread over a wide range. In other words uncertainty of velocity is large. The wave function is represented in literature by the symbol ψ (r, t). The rate at which the wave function changes with time is embodied in what is known as ***Schrödinger's Equation***, named after the well known Austrian physicist Erwin Schrödinger. The time dependant general Schrödinger equation is given by,

$$i \cdot \hbar \frac{\partial}{\partial t}[\psi(r,t)] = H\psi(r,t)$$

where, "i" represents the imaginary number i = $\sqrt{-1}$, "$\hbar$" is the reduced Planck's constant given by $\hbar = \frac{h}{2\pi}$, the symbol $\partial/\partial t$ indicates the partial derivative with respect to time, ψ (the Greek letter psi) is the wave function of the quantum system, "r" and "t" are the position vector and time variable respectively and "H" is the Hamiltonian operator which characterizes the total energy of system wave function.

If we know the wave function, at any given time, we can use the Schrödinger equation to calculate the wave function at any other time, past or future. Instead of predicting both position and velocity, we can only predict the time evolution of the wave function. This can allow us to predict either position or velocity, but not both accurately. While "Ψ" has no physical interpretation, the square of its absolute magnitude, namely $|\psi|^2$, evaluated at a particular place and at a particular time, is proportional to the probability of finding the body at that place and time. A deeper interpretation concerning Schrödinger's equation is quite complex and is not attempted here as it falls outside the scope of this chapter.

11.7 FEYNMAN'S PATH INTEGRAL FORMULATION

In 1948, Richard Feynman an exceptionally gifted physicist proposed a powerful new way of conceptualizing quantum mechanics, for which Feynman along with Julian Schwinger and Sin Itiro Tomanaga were jointly awarded the Nobel Prize for physics in 1965. Feynman challenged the basic classical assumption that each particle has a particular history as it traveled from position A at time t_1 to position B at time t_2. Instead he suggested that the particle travels between A and B along every possible path through space time. With each trajectory, Feynman associated two numbers, namely amplitude and phase. The probability of a particle going from A to B is found by adding up the waves associated with every possible path through A and B. However, in every day world we are accustomed to the particle following a single path between the origin and the destination. This is in agreement with Feynman's multiple history or sum over histories idea. This is because for large objects, his rule for assigning numbers for each path ensures that all paths but one cancel out, when their contributions are combined. As far as macroscopic paths are concerned, only one out of the infinity of paths matter. The concept of path integral formulation is illustrated in Fig 1 below.

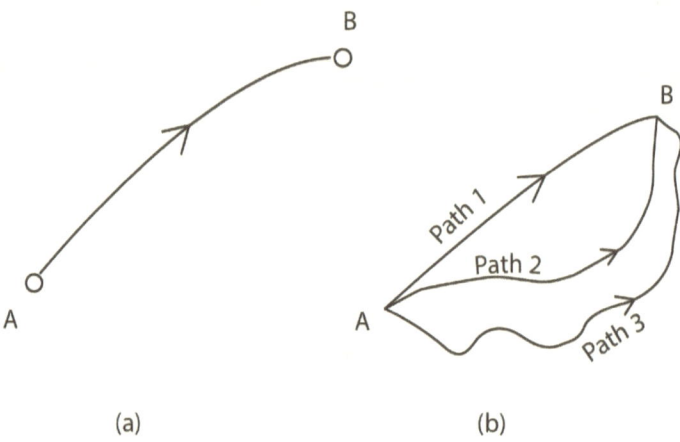

a) Classical path of the particle moving from A to B

b) Three of the possibly many paths that contribute to the quantum amplitude of the particle moving from A to B

Fig 1: Feynman's Path Integral

To sum up, relativity and quantum mechanics represent two entirely diverse streams of thought, each operating extremely well in their own respective domains. To unify both these profound theory into one single unified theory is the challenge which confronts physicists today. Whether they will succeed or not in the near future, remains to be seen.

Chapter 12

BIOLOGY'S BIG BANG

"Once we accept our limits, we go beyond them"
— Albert Einstein

12.1 EARLY DEVELOPMENTS

Biology is essentially based on observation, description and classification. It is in this context that evolutionary biology plays a fundamental role. Perhaps no single individual has exerted such a sweeping influence on evolutionary biology as Charles Darwin (1809-1882) did. His enduring legacy covered the whole range of evolution from natural selection to the origin of man. Some of his seminal contributions are contained in his two books namely, "The Origin of Species"- (1859) and the "Descent of Man"- (1871). In his first book, Darwin stated his ideas on evolution by natural selection. Species mutation is led by individuals whose inherited characteristics enabled them to dominate their habitat. In this connection he wrote "In the struggle for survival, the fittest win at the expense of their rivals because they succeed in adapting themselves best to their environment". In his second book Darwin applied evolutionary theory to human beings. He hypothesized that Homo sapiens and apes descend from a common ancestor. He asserted that "Man still bears in his bodily frame the indelible stamp of his lowly origin". In one stroke Darwin established that humans are just a species of animals rather than a lordly creature created by God.

Simultaneously, the biologist Gregor Mendel (1822-1884) did pioneering study in the modern science of genetics, a work for which he was posthumously recognized. He established many of the rules of heredity, which are often referred to as the Mendelian Law of Inheritance. His results were further extended by J.B.S Haldane (1892-1964). Haldane unified Darwinian evolution by natural selection and Mendelian genetics. A polymath in his own right, he was the first to suggest the feasibility of in-vitro fertilization. While nineteenth century work in biology concentrated on evolution and genetics, twentieth century saw the emergence of a new discipline called molecular biology.

12.2 INTRODUCTION TO MOLECULAR BIOLOGY

Molecular biology got a big push when in 1953 two biologists Francis Crick and James Watson unveiled the double helix structure of the DNA. Their discovery was in no small measure assisted by the X-ray crystallographic studies of the DNA made by a young biologist Rosalind Franklin. Unfortunately her early death at the young age of thirty seven deprived her of the Nobel Prize in Medicine, which was later awarded jointly to Watson, Crick and Maurice Wilkins in 1962.

DNA (Deoxyribo Nucleic Acid) is found in the heart of every living cell. It carries all the instructions for an organization to build, maintain and replicate itself. By replicating and passing their DNA, animals, plants and even micro-organisms can impart their characteristics to their offspring. In humans half the DNA in our cells is contributed by the mother and the other half by the father. Thus we inherit a mixture of characteristics from our parents. DNA represents a very long genetic code and this code is unique for every living being. Our understanding of the DNA has revolutionized the entire discipline of biology. In fact, it confirms and further refines Darwin's theory of evolution.

The DNA structure looks like a very long twisted ladder known as the ***Double Helix***. On either backbone of the ladder are located in sequence, chemical building blocks known as ***Nucleotides***. Four different types of nucleotides have been identified. These are denoted by the letters **A, C, G** and **T. A** stands for ***Adenine***, **C** for ***Cytosine***, **G** for ***Guanine*** and **T** for ***Thymine***. The exact order, in which the four letters **A, C, G**, and **T** are arranged along the backbone of the double helix ladder, varies resulting in an extremely long code. The nucleotides or bases on either side of the backbone of the ladder are linked by rungs of the ladder subject to certain restrictions. The base A can only link with base T and the base C can only link with base G. This ***base pairing principle*** is extremely important for proper DNA functioning. Human DNA has about 6 billion rungs on the ladder. The complex code of A C G T arranged along the backbone of the ladder is different for every person, except in the case of identical twins. The complete DNA sequence of any organism is called the ***Genome***.

The study of how just four letters are able to form a unique code stretching over six billion bases is extremely complex. An organization inside each cell reads three consecutive letters of the DNA thus forming successive blocks. Each such block is called a ***Codon***. These blocks may contain bases, such as ACG, CAG, ACT etc. It is easy to see that if one chooses 3 letters out of 4 letters, there exist 64 different possible combinations. But out of these combinations only 20 are utilized to codify the so called ***Amino acids***. Normally, amino acids are codified by one codon, but some amino acids use more than one codon for codification. For example, the amino acid Lysine is coded by two codons, namely AAA and AAG. There are a number of codons which do not code for any amino acid. They either serve as signals, to mark the end of a signal coding sequence or for some other purpose. Those that mark the end of a sequence are known as ***stop codons***. The amino acids can be joined together in an incredible number of combinations to deliver an enormous number of proteins. In this context a ***gene*** is a stretch of DNA that encodes a protein. *All proteins are encoded by genes but not all genes encode a protein.* The following flow chart illustrates the transition from a nucleotide to a protein.

Fig 1: Transition from nucleotides to proteins

Different combinations of amino acids lead to different types of proteins with widely varying functions. For example, their functions range from the production of hormones (like testosterone and estrogen) to the production of important molecules that help to form hair, skin, muscle etc. There are hundreds of thousands of different proteins within the human body. Variation in our genes causes variations in proteins in our cells, which in turn lead to different human characteristics. A common misconception is that one gene is solely associated with one particular trait of an organism. In fact, physical traits arise from a combination of many genes. It is now possible to investigate what a particular gene does by either removing it or by changing the sequence of base pairs. Mapping of the human genome known as the *human genome project* was one of the principal tasks successfully completed by a body of international scientists in 2003. According to one assessment, the number of protein-coding genes in human genome range between 20,000 and 25,000. A comparison in this respect with a microscopic worm technically referred to as C. Elegans is revealing. It has much the same organs as higher animals have such as gut, mouth, etc, but it hardly consists of 1000 cells compared with over 50 trillion or so cells in the human body. Further the C. Elegans has about the same number of genes (about 20,000) as the human body. Also in this case, the percentage of the genome that does not code for proteins is 75%, as compared to an abnormally high 98% in the case of human beings. This shows how things change drastically as we go up the evolutionary tree. Clearly, the number of genes in an organism gives no indication about the complexity of the organism.

12.3 REPLICATION OF THE DNA

The double helix structure of the DNA, helps us to understand how DNA replicates. To begin with, making use of the chemicals in the cell, DNA untwists itself and then the two strands split down the middle, analogous to a zip in action. Because of the base pairing property of DNA, A always pairs with T and C always pairs with G. Thus, when a single cell divides itself to form two daughter cells, each daughter cell is an exact template of the original cell. The whole process starts with a ***single starter cell*** also known as the ***Zygote***, which is formed when a sperm fuses with an egg in the uterus of the mother.

The starter cell contains 6 billion base pairs- half inherited from the father and the other half from the mother. This cell division process continues during the early stages of division, but at some stage an internal mechanism within the cell regulates the gene expression resulting in the creation of specific cells which are identified with different parts of the body. This process is known as ***differentiation***. A differentiated cell for example, a liver cell is different from an undifferentiated cell such as the one associated with a zygote. The undifferentiated cells are called ***pluripotent cells***. All embryonic cells are pluripotent. The need for cell division is obvious. As the body grows, more and more cells are required for taking care of its growth and also to replenish the dead cells generated due to the growth process.

12.4 STRUCTURE OF CHROMOSOMES

The genome sequences in animals and plants are so long that it cannot be accommodated as such in a single cell. For example in a human being, the length of a single genome accommodating 6 billion bases will be about 3 meters long. However, the cell which is to accommodate this strand has hardly a length of 2 to 3 microns. For this reason and also to provide easy accessibility to specific regions of DNA, the entire genome is tightly packed into X shaped bundles known as ***chromosomes***. In humans the genetic code is distributed among 23 sets of chromosomes. Each set consists of 2 chromosomes - one from the father and the other from the mother. Hence the human body contains in all 46 chromosomes in each cell. In a normal person, the first 22 sets of chromosomes look very much alike and are known as ***autosomes***. The 23rd set is what determines the sex of the child when the zygote is formed and is known as the ***sex chromosome***. In a female there are two similar sex chromosomes in the 23rd set and each are labeled as X- chromosome. For a male, in the 23rd set, one chromosome is similar to the X-chromosome in the female while the other chromosome known as the Y- chromosome is comparatively very small. The structure of the male and female chromosomes is illustrated in Fig 2.

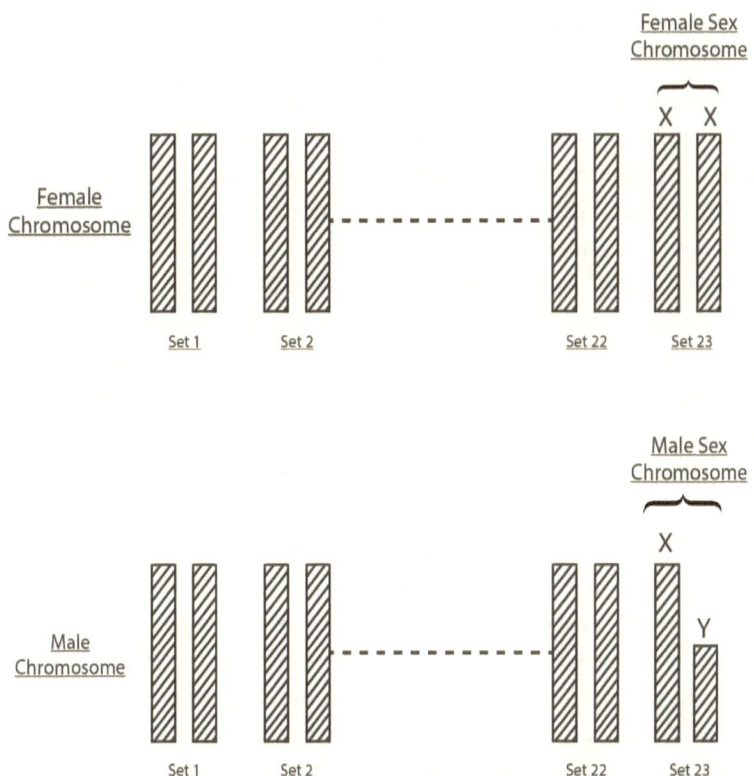

Fig 2: Chromosome structure in a human cell

When a sperm fuses with an egg to produce an embryo, the following possibilities arise. In the case of the female child, apart from the first 22 sets of non-sex chromosomes which are equally shared by the parents, the sex chromosomes (set 23) contributes one X chromosome from the mother and another X chromosome from the father. In the case of a male child, the mother contributes X chromosome while the father contributes Y chromosome. The X chromosome codes for about 1300 protein coding genes while the Y chromosome codes only for 40 to 50 protein coding genes. In persons suffering from a disorder known as Down's syndrome, there are three chromosomes in each of the 21 sets, thus having 63 chromosomes (21 x 3) in all, instead of the usual (23 x 2), i.e. 46 chromosomes as in a normal child. This is a major abnormality in individuals who are born with this disorder.

The ends of individual chromosomes have a protective covering known as **_Telomeres_**. We may compare it to the little plastic caps at the end of a shoe lace which prevents fraying of the lace at both ends. Telomeres play a similar role. As the cell divides, the telomeres wear down and it ceases to protect the ends of the chromosomes. This results in malfunctioning of the cells. Several scientific studies have established the link between telomere length and longevity/disease of the individual. It is found that people with longer telomere lengths live longer, while those with shorter ones are linked with chronic disease and early mortality. The wearing down or shortening of the telomeres with age is a perfectly natural process. However, if the shortening is abnormal it may be delayed by external intervention. These include proper stress management, healthy food habits, good sleep routine and regular daily exercise. Three scientists namely, Elizabeth H. Blackburn (University of California in San Francisco), Carol W. Greider (John Hopkins University School of Medicine) and Jack W. Szostak (Massachusetts General Hospital at Boston) who have done extensive work on Telomeres and the enzyme *Telomerase* associated with it, were awarded the Nobel Prize in Medicine in 2009. Dr Elizabeth Blackburn is also a co-author of an authoritative book on the subject titled "Telomere effect". The discoveries in this area have added a new dimension to our understanding of the working of the cell.

12.5 MESSENGER RNA (mRNA) AND RIBOSOME

How does the DNA in the chromosomes act as a script for producing proteins? This is achieved by an intermediary protein molecule called the messenger RNA (mRNA), where RNA stands for Ribo Nucleic Acid. mRNA is very much like the DNA except that it has a single strand as opposed to two strands structure in DNA. When a particular stretch of DNA is "read" so that a protein can be produced using a stretch of the script, a complex set of proteins inside the cell, unzips the right piece of DNA and makes authentic mRNA copies of it. These mRNA copies are utilized as templates by structures within the cell, where the 3 letter Codon sequence is read and then stitched together to form the right amino acids. This in turn ultimately leads to longer protein chains. Out of the 6 billion base pairs contained in each cell of our body, only 120 million base pairs code for the proteins. This constitutes only 2% of the total human genome. To summarize, what the mRNA does is to make copies of the salient portions of the DNA as required and send them on to protein manufacturing centers of different cells.

For decades, major bio-technological companies have invested large amount of money and resources in research and development and made significant advances in micro-biology and genetic engineering, with DNA (the underlying code or blue print) and the proteins (the functional molecules within the cell), being the focal points of their research. The investigation and study of RNA was accorded lesser importance till a few years back. This was because, its role in cell metabolism was considered relatively unimportant, as it acted only as a carrier of information from the DNA to the protein factory within the cell. Clearly, the proteins are extremely important in carrying out a multiplicity of functions that enables life to carry on. Even small alterations in protein coding genes can lead to mutations with consequent devastating effects. In this context the so called *Ribosome* translates the passive DNA information conveyed through mRNA into form and substance. The Ribosome's role is that of a translator which translates genetic information into action (protein manufacture). Ribosome exists in all cells of all living organisms from bacteria to human beings. As no creatures can survive without ribosomes, they serve as perfect targets for drugs. Many of the antibodies which are created in our bodies due to the intake of drugs, attack only the bacteria's ribosomes, but leave human ribosome's untouched.

Ribosomes produce proteins in the *cytoplasm*, the liquid inside the cell which has a nucleus at its core. The messenger RNA, which copies the genetic message and which lingers outside the nucleus, is captured by the ribosome. The ribosome in turn reads the information conveyed by the messenger RNA and based on this information produces proteins. Scientists refer to this action as *Translation*. To understand how ribosome molecule functions, it is important to determine its structure. Ribosomes are hardly 25 nano meters in size. Its complex atomic structure using X-ray crystallographic methods was revealed by three scientists namely, Adae E. Yonath (Weizmann Institute of Science, in Reovot, Israel), Thomas A. Steitz (Yale University, New Haven, USA) and Venkatraman Ramakrishnan (MRC Laboratory, Cambridge, UK). For their outstanding work they were jointly awarded the Nobel Prize in Chemistry in 2009. Ribosomes play a central role in developing potentially new therapies which are of benefit to humanity. Its discovery has broadened our understanding of the human cell and shed light on disease control mechanisms.

12.6 THE IDEA OF EPIGENETICS

Epigenetics refers to the study of non-inherited changes in the gene function that do not involve changes in DNA. Any process that alters gene activity without changing the DNA sequence comes within its ambit. The effect of epigenetics can be simply explained by the following example. Consider two identical twins that have the same genetic code. They share the same mother's womb but on birth they are separated and brought up in different environments. In some cases, while one of the twins appears to be completely healthy, the other twin has serious health problems. One has to presume in this case that gene activity is controlled by the environment resulting in the different states of health of the twins. The question then arises how environment can have different impacts on two genetically identical persons. The age old problem underlying this question regarding "Nature versus Nurture" is yet to be satisfactorily resolved. Probably, the science of epigenetics may provide an answer to this complex problem.

Chapter 13

SYNTHETIC BIOLOGY – PLAYING GOD!

*"It is a riddle, wrapped in a mystery inside an enigma,
but perhaps there is a key"*
— Wintston Churchill

13.1 INTRODUCTION

Exciting times lie ahead for biologists and geneticists. The past six decades have witnessed some path breaking developments in this discipline, setting the tone for future progress in many new directions. A few of these developments are listed below:

1. The double helix structure of DNA developed by molecular biologists Crick and Watson in 1953.
2. Birth of the first In-Vitro Fertilization (IVF) baby, Louis Brown in 1978.
3. Cloning of Dolly the sheep in 1997.
4. Complete sequencing of human genome by 2003.
5. Gene therapy and gene editing.

At this current rate of progress many believe that the pre-eminent position occupied by Physics in the 20th century, will be replaced by Biology and related disciplines during the 21st century

13.2 RECAPITULATION OF SOME BASIC CONCEPTS

The code of life among all living organisms, starting from the lowly bacteria, to highly evolved human beings is enshrined in a chemical called DNA (deoxyribonucleic acid). The back bone or spine of DNA represents a chain of sugars and phosphates strung together. If DNA represents a code, it has to be identified by symbols that can be read. These symbols are denoted by 4 letters A, C, G and T. "A" represents *Adenine*, "C" represents *Cytosine*, "G" represents *Guanine* and "T" represents *Thymine*. These 4 letters are also referred to as bases or nucleotides of the DNA. The complex 3 dimensional structure of DNA molecule was first unraveled in 1953, by Nobel laureates Francis Crick and James Watson. Through crystallographic studies, they arrived at a double helix structure analogous to a twisted ladder with rungs all along the ladder. The bases A, C, G, T are distributed along the two backbones of the ladder, in no particular order with parallel rungs connecting them. The connections strictly follow what is known as the ***base pairing principle***. Thus

"A" in the backbone 1 is only connected to "T" of backbone 2 and "C" of back bone 1 is only connected to "G" of back bone 2. There are approximately 3 billion base pairs on this ladder as far as human beings are concerned. The entire sequence of bases associated with each of these backbones is known as the ***genome*** of the organism. The genome of every individual is unique in the sense that no two individuals carry an identical gene sequence, except in cases where both the individuals happen to be ***identical twins***. The code adopted for deciphering the DNA comprises of a triplet of bases, called blocks and some of these blocks are identified as amino acids. With 4 bases to choose from and with any three of them forming a block, their combination reduces to 64 distinct triplets, out of which 20 triplets are all that are needed to identify 20 and odd amino acids. These triplets in general are known as ***codons.*** The codons are strung together to form a variety of combinations of DNA some of which are identified as ***genes***. Some of these genes are associated with hundreds of proteins manufactured in the different cells of the human body, such as neurons, skin, liver, muscles, kidney etc. It is important to note that all proteins are encoded by genes. However, the reverse is not always true. There are thousands of genes which do not encode any protein. The protein coding genes do not often follow a continuous sequence. There are often breaks in the gene code known as ***introns*** which either serves as ***stops*** or for signaling the beginning and end of a protein coding gene sequence.

The real question is how exactly does a succession of codons in a human genome act as scripts for providing proteins as and when required? This is achieved via intermediary molecules called RNA (ribonucleic acid). It is very much like the DNA except for two significant differences. Its 4 bases are denoted by letters A, C, G and U, unlike A, C, G, and T for DNA. The T in DNA is replaced by U in RNA, where U stands for ***Uracil***. Further, RNA is a single stranded structure unlike its counterpart namely, DNA which is double stranded.

A certain class of RNA known as ***messenger RNA*** (mRNA) conveys a message from the DNA to the protein factory inside the cell to produce proteins, when commanded to do so by the so called ***regulatory genes*** inside the cell. The mRNA copies stretches of protein-encoding genes from the original, by using the *base pairing principle*. The mRNA then delivers the message to special RNA structures called ***Ribosomes*** inside the cell. The ribosomes serve to translate the message to give it form and substance culminating in the manufacture of different types of proteins, enzymes, hormones etc.

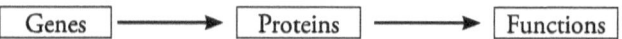

As already stated human genome contains 6 billion base pairs out of which protein coding genes occupy hardly 2% of the entire stretch. A small fraction of the remaining part of the genome is utilized in forming ***regulatory genes*** and for the creation of RNA molecules.

Even after taking all of this into account, there are still vast stretches of DNA which apparently do not seem to have any role to play. They are often referred to as *junk DNA*.

In this context, it is interesting to compare the human genome with the genome of other living organisms. At one extreme we have the single celled bacteria with 90% of its genome utilized for manufacturing proteins. This figure falls sharply in the case of the miniscule soil dwelling worm known as *C*-elegans which utilizes 25% of its genome for protein coding. Further up the evolutionary tree, we have humans who utilize just 2% of their genome for protein coding purposes. This is surprising, as the more complex the organism one would expect more percentage of the genome to account for protein coding. Strangely, both *C*-elegans and humans have approximately the same number of genes (about 20,000) involved in the manufacture of proteins. This clearly demonstrates that the actual number of genes is of far less importance as compared to the way the organisms put them to effective use.

The genome sequence of humans stretches for a comparatively long length and some compressive arrangement is required to contain them within the nucleus of a cell which occupies hardly 2 to 3 cubic microns in volume (1 micron= 10^{-6} meter). For this reason and also to facilitate easy accessibility, the genome is tightly packed in X-shaped bundles known as *chromosomes*. They are tucked inside the nucleus of the cell which comprises of 46 chromosomes. In humans the genome is made up of 46 chromosomes with equal contributions made by the father and mother of the individual. These 46 chromosomes are arranged in 23 pairs. The first 22 pairs are similar in size in the case of both males and females. The difference lies in the 23rd pair known as the *sex chromosome*. Here the female has two similar chromosomes known as X-chromosomes, whereas the male has one X-chromosome and one Y- chromosome. Compared to the X-chromosome, the Y-chromosome is considerably smaller. The former accommodates about 1000 genes, whereas as the latter hardly accounts for about 27 genes.

When the sperm fertilizes an egg to form an embryo, the following possibilities arise. In the case of a female child, the mother contributes an X-chromosome and so does the father. However, in the case of a male child the mother contributes the X-chromosome whereas the father contributes Y-chromosome. Thus all females are born with XX sex chromosomes whereas all males have XY chromosomes.

13.3 REPLICATION, REGULATION AND RECOMBINATION OF GENES

The process of life starts with a single celled embryo, born out of the fertilization of an egg and a sperm. The genetic material of the embryo is shared equally by the father and mother of the future child. Each and every cell in our body houses a nucleus with the genome packed in 23 pairs of chromosomes. The nucleus is immersed in a thick liquid inside the cell called the *cytoplasm*. The cellular material of the embryo comes exclusively from the egg. The sperm merely serves as delivery conduit for the contribution of its share of the DNA. Apart from proteins, ribosomes, nutrients etc, the cell also houses a specialized structure called **mitochondria**. These are the energy producing factories of the cell. It carries a small independent genome inside it (not in the cell nucleus). All individuals male or female inherit the mitochondria genes from their mothers side and the fathers have no part to play in their propagation across generations.

To summarize the human body is a strange concoction of DNA, RNA, and Proteins. Other ingredients involved are sugars and fats (too abundant in the case of some people), water and millions of friendly microbes. Our body contains 50 trillion to 100 trillion cells. Human genome houses 3 billion base pairs of DNA and 23 pairs of chromosomes. From fending off bacteria to digestion of food, there are a variety of processes occuring in our cells.

The Watson- Crick base pairing of the double helix structure facilitates the carrying out of three remarkable activities in the genes- namely, **replication, regulation and recombination**.

Cells are being constantly replicated in our body to take care of dead cells or cells lost due to injury. **Replication** takes place as follows. The double helix is first split right in the middle and each part is used as a template to create its unique counter part. Thus we get two identical cells out of the first cell and these two multiply into four cells etc, and the process goes on. However, if this replication process goes on unchecked, then it may lead to diseases like cancer.

Regulation works through transcription of DNA into RNA. Thus a strand of DNA is used to convey a message through mRNA to the protein factory in the cell. Here again it is the gene pairing principle that allows a gene to be copied into a single strand RNA. In the **recombination** process also, the strategy of matching base against base is deployed in the restoration of damaged DNA. The damaged copy of a gene can thus be reconstructed using the complementary strand or the second copy of DNA as its guide.

13.4 STEM CELLS, EMBRYONIC CELLS AND INDUCED PLURIPOTENT STEM CELLS (IPSC)

A *stem cell* is a unique kind of cell that is distinguished by two properties.

1. It can renew itself – i.e. it can give rise to more stem cells, which in turn can differentiate to form the functional cells of any organ in an organism.
2. A stem cell can give rise to other functional cell types such as nerve cells through differentiation.

Most stem cells reside in particular organs and tissues and only give rise to limited repertoire of cells. For example, stem cells from bone marrow produce only red blood cells and not liver cells. ***Embryonic stem cells*** or *E* s cells reside in the inner sheath of the human embryo and are extremely versatile. They can give rise by the so called differentiation process to every other type of cell in the organism- blood, brain, intestine, muscle, bone, skin tissue etc. This property of the embryonic cell capable of differentiating to form any other type of cell is referred to as ***pluripotent***. Embryonic cells have another useful property namely; it can be removed from the embryo of an organism at a very early stage and grown in petri dishes in a laboratory. These cells can grow continuously in a culture medium. They can be frozen in vials and thawed back into life whenever required. Genes can be inserted into their genome or removed from their genome with relative ease. However, this type of modification of embryonic stem cells *in-vitro* and then re-implanting them into the womb faces many obstacles. First, many countries have disallowed using E s cells for gene modification followed by its implantation in the womb, primarily from ethical considerations. Secondly, the problem of matching such cells to avoid rejection is very real. The question then revolves around how to obtain E s cells from adult stem cells, if possible, adopting simple procedures in the laboratory?

In 2006, a Japanese professor Shinya Yamanaka showed that not only is this transformation possible, but can be achieved using simple procedures. Prior to this, in the 1960's the British biologist John Gurdon attempted to solve a slightly different but related problem. He first used an unfertilized frog's egg and immobilized its nucleus by exposing it to a ray of ultraviolet light. He then injected into it, the nucleus of an adult frog cell. In proper time a functional tadpole arrived – a perfect clone of the original donor frog. However, the success rate of this procedure was abysmally poor. Juxtaposed against this background, Yamanaka's work is highly significant. Like Gurdon, he also started by attempting to preserve an adult cell's identity by using the stem cells from an adult mouse's skin. The idea was to create pluripotent cells out of differentiated cells in the laboratory. After years of experiment, he found that by introducing the proteins transcripted by just 4 specific genes, he could transform a part of the mature stem cells into pluripotent cells. Yamanaka called the cells that he created ***induced pluripotent stem cells*** with the acronym iPSC. The iPS cells thus showed a new way forward. All that one has to do is to take a small scraping of skin cells from the patient, and grow these cells in a culture using the four so called Yamanaka reagents to create a large number of iPS cells. Next, if one decide for example to turn these cells into, beta (β) cells via differentiation (in case of insulin deficiency) and transfer them back through the patient, there won't be any rejection since the patient is using his own cells. However, there is a problem. One of Yamanaka's factors (reagents), is known to actively promote cancer. Notwithstanding this setback, work is in progress to find a safe alternative for this particular reagent. For their seminal discovery that mature cells could be reprogrammed to become pluripotent cells, Shinya Yamanaka and John Gurdon were jointly awarded the Nobel Prize in Medicine for the year 2012.

13.5 CLONING AND GENE THERAPY

The next landmark achievement in molecular biology was reported in 1997 when British embryologists Keith Cambell and Ian Wilmut at the Roslin Institute in Scotland, created the first mammalian clone known as Dolly the sheep. Like Gurdon who preceded them, they adopted the SCNT (Somatic Cell Nuclear Transfer) procedure for cloning. They first transferred the nucleus from a cell in the mammary gland of an adult sheep called Molly into the unfertilized egg of another sheep from whose egg its original nucleus had been removed. Next they transplanted this egg into the uterus of a recipient sheep which however merely served as a surrogate mother. After many such unsuccessful trials, they finally succeeded in producing a sheep named Dolly which was a clone or an identical replica in every respect of the original sheep Molly which they attempted to clone. The next question that arose was that if one could clone frogs and sheep, why not human beings? Technically and theoretically this is possible; however one should not forget that merely producing a clone with identical genes does not make two persons identical in all respects and there are many other factors like environment, education, experience etc, to be taken into consideration. These factors cannot be replicated between two individuals. Another deterrent for human cloning is that it is prohibited by law in most countries on ethical grounds.

13.6 GENE THERAPY

The completion of the epoch making Human Genome Project in 2003 was a landmark achievements and opened up new avenues of research in gene therapy. It made it possible for us to read the 3 billion and odd pairs of nucleotides in the DNA chain of every human being. We are now in a position to identify those genes from among the thousands of genes whose mutations could cause deadly and debilitating disorders such as Cystic fibrosis and Sickle celled anemia. Today, *prenatal* genetic testing could reveal whether an embryo is affected because of some defective gene. There are two options here for further action. Either, the couple can decide on going in for an abortion at a sufficiently early stage of the pregnancy or the genetically affected embryo could be removed and cultured in a laboratory to rectify the mutation. The nucleus of the rectified cell could then be placed in an un-nucleated egg cell to eventually produce a healthy baby.

How genes and gene therapy work: We recall that a gene represents a collection of bases distributed randomly all along the human genome. Some of these genes serve as blue prints for making specific proteins. In fact genes spell out a sequence of amino acids some of which may lead to manufacture of proteins. All cells in the human body as already stated, carry the same genome in their nucleus. However, if we consider specific cells, a liver cell for example, functions differently from a brain cell or a kidney cell. This is because different types of cells use different sets of genes to produce proteins specifically suited for that particular type of cell. To state it differently, each cell copies only selected genes from the entire genome available in its nucleus via the so-called messenger RNA to transcript the specific protein required. This mechanism serves as a template from which the proteins are constructed and manufactured inside the cell, as required.

In the case of a mutated gene, the protein produced will not serve the purpose for which it is intended and in some cases may even work aggressively. This in turn may lead to symptoms of diseases. In a simple version of gene therapy, healthy genes are directly injected into the selected organ to offset gene mutation. Genes are currently provided to patients in two basic ways. In both cases, the genes are first put into transporters or vectors capable of transporting genes to the targeted cells. In this procedure, scientists first remove cells from a selected tissue in a patient, genetically correct it and then expose it to gene transfer vectors in the laboratory. The genetically corrected cell is then returned to the individual via the chosen vector which carries it. At other times, the corrected gene is introduced directly into the body, generally into the tissue to be treated. In either case, the main aim is to get the corrected gene delivered at the desired sites.

The key problem in gene therapy is to choose a vector which is a safe and effective gene delivery vehicle. It is in this connection that viruses are chosen as potential vectors to perform this task. Viruses are self replicating genes wrapped in a protein coat – *"a piece of bad news wrapped in a protein coat"* as famously remarked by the noted biologist and Nobel laureate Peter Medawar. They replicate mostly inside living cells of other organisms. Unlike bacteria virus possess no nucleus and have no cell. In theory at least a modified and tamed virus (one that has been rendered partially harmless in the host's body), should be capable of transferring the healthy or corrected genes inside the cells of various body parts of an organism without excessive multiplication and without acting as harbingers of any disease. In this context, it is relevant to explore the roles played by bacteria and virus inside a cell. Bacteria and fungus enter a cell as immigrants, to use an analogy. Some are friendly and some are pathogenic- i.e they are harmful and disease causing. They are subsequently rendered harmless by the action of antibiotics administered.

The role of the viruses is quite different. To continue with the immigrant analogy, the viruses on immigration insert their genetic material into the host's genome and reside there for an indefinite period. In the case of an HIV virus- which is commonly chosen as a vector to deliver the healthy genes – the action is slightly different. Its genetic material is composed of the single stranded RNA and not double stranded DNA, as in the case of normal viruses. HIV virus integrates into its host genome by first making a copy of its RNA (known as reverse transcriptase) into DNA and then integrating itself into its host's genome. Because of this reverse transcriptase, the HIV viruses come under the category of retro-viruses. In such situations, the ideal action for the host to take is to generate an enzyme that can first recognize the foreign DNA and then take corrective action by chopping it off. But, unfortunately this does not often work because some of the retroviruses have the uncanny ability of changing its DNA composition over a period, thus rendering the action of the enzyme ineffective.

The viruses most extensively considered for gene therapy are the so called ***retro-viruses***. They splice copies of their genes, permanently into chromosomes of the cells that they invade. Such copies are passed on for future replications of those cells during the host's life time. In contrast, a special kind of viruses (non-retroviruses), do not integrate their genetic material into its host's chromosome. Their genes generally function for a transitory period in the cells of the host's body before they are overcome by the immune system. Among the retro-viruses that are most commonly used, are a sub-set of tamed HIV viruses. However, among the ordinary viruses the most commonly employed, include the Adeno virus, which do not cause any serious problem through proliferation.

In spite of its appeal for being used as a vector, retro-viruses suffer from two major setbacks. Their genes integrate randomly into the host's genes and may sometimes cause serious disruption in organ functioning. Further, they fail to transfer healthy genes into cell types that cannot divide for example, mature neuron cells and skeletal muscle cells. This is because viruses can reach chromosomes only where the membrane surrounding the nucleus of the host cell dissolves – and this can occur only during cell division. The disadvantage of Adeno viruses is that they come under severe attack from the body's immune system which sometimes results in the destruction of the virus along with the corrected gene that it is supposed to carry. Once the immune system is alerted of the presence of any virus, it tries to eliminate it quickly, in case they are delivered a second time.

13.7 GENETICALLY LINKED DISEASES

There are various categories of genetically linked diseases. On the one hand we have instances where the mutation of a single gene in the genome of an individual invariably leads to the disease with which it is associated and that too with almost hundred per cent certainty.

Diseases such as Cystic fibrosis and Hemophilia belong to this category. Cystic fibrosis is a devastating disease affecting multiple organs like lungs, pancreas, bile duct and intestines. Its only cure lies in the correction of a single gene which has undergone mutation. Hemophilia results in the failure of the blood to clot. It is due to the lack of a functional clotting protein in the blood. It is one of the first genetic diseases associated with a single gene. Further, it is also hereditary. Gene therapists have now succeeded in synthetically making the blood clotting protein and injecting this directly into the blood stream of the persons affected. This will work as long as its effect lasts. If this approach works, other genetic diseases caused by a single gene mutation will be vastly more treatable than it was previously thought.

In yet another category of genetically transmitted diseases, the mutation of several genes leads to a single disease. The occurrence of breast cancer is a classical example. It is now well established that breast and ovarian cancer has a family history. It is caused by the mutation of a gene called BRCA1. However, this mutation alone need not necessarily result in breast cancer, though the probability of occurrence is high. It is the simultaneous occurrence of other mutated genes, distributed along the DNA chain, along with BRCA1 mutation which is often responsible for breast cancer.

Another example of mutation of multiple genes causing a single genetically transmitted disease, relates to what is known as the Down syndrome. In this case the patient has three copies of chromosome 21, instead of only two as in the normal case. This extra chromosome contains 300 and odd genes strung together. Men and women born with this abnormality have heart diseases, increasing risk of blood cancers and cognitive deficiency. Other genetically transmitted diseases like diabetes, hypertension, schizophrenia etc, are usually caused by multiple genes distributed diffusely throughout the genome. In many cases the genetic component of a particular disease only acts as a trigger along with the other environmental variables such as diet, age, smoking, nutrition etc, for the onset of the disease.

Gene therapy has progressed a lot since the completion of the human genome project in 2003. It is now possible to obtain the entire DNA sequence of any individual on payment of a moderate sum of $1000. This is a one time investment and the test can be done starting from the moment the child is born. The entire genome can then be stored in a small smart card and use the readily available data as and when required. We recall here the recent well publicized case of a Hollywood actress who after finding that her genome revealed the mutation of BRAC1 gene (which indirectly leads to breast cancer) and also taking into account her family history, went in for a double mastectomy as a precautionary measure.

13.8 DNA EDITING TECHNIQUES – CRISPR- CAS9 AND BEYOND

For a long time, it was the dream of microbiologists to intentionally change the human genome in order to get rid of their aberrations which cause genetically transmitted diseases. Gene delivery methods such as the one described earlier using viruses as a vector to correct gene mutations in general, proved to be highly unreliable. At worst, it turned out to be extremely dangerous if the viruses missed its target cells.

Now, things have changed for the better. Novel gene editing techniques allow geneticists to make considerable alterations in the human genome with ease and precision. In principle a single letter of DNA can be substituted by another letter leaving the remaining 3 billion base pairs of the DNA largely untouched. All this has been made possible using a new technique called CRISPR-Cas9 discovered as late as 2012 by two molecular biologists Jennifer Doudna and Emmanelle Charpentier of USA. They got their inspiration by observing how bacteria effectively fought back the virus trying to invade and destroy it. Viruses in general, have evolved genetic mechanisms to invade and kill bacteria. The bacteria in turn have countered it by the creation of genes to fight back the viruses. In this context, scientists found that certain sections of DNA were frequently repeated in the bacterial genome attacked by a virus- These sequences are known as ***Clustered Regularly Interspaced Short Palindromic Repeats***, in short ***CRISPR***. After a virus invades a bacterial cell, enzymes generated by the bacteria cut and paste viral genomic material between CRISPR sequences. This leaves a genetic memory for an ***RNA guide***, which in turn uses an enzyme called ***Cas9*** to destroy the viral DNA, should the invader return to invade the cell. What the researchers found in 2012 was that in the case of humans, an RNA guide could be encrypted to target any desired DNA sequence in the human genome.

CRISPR-Cas9 has essentially two components namely, a guide RNA that finds and binds itself to the desired location in the genome and Cas9 that cuts the DNA at that location. When the gene tries to repair the cut, errors arise in the sequences that cause mutations, thus disabling the gene. This method can in principle be extended enabling any gene or a set of genes to be changed in an intentional manner and new set of genes to be substituted in its place. At present CRISPR-Cas9 works more precisely, more powerfully and more efficiently than any other genome altering method. It must be noted that CRISPR is merely a tool and before using it, one must have an idea of which gene has to turn - "on" or "off". CRISPR is a method of such great consequence that the prestigious US magazine *Science*, listed it as one of the major break through tools for the year 2015.

However, many challenges lie ahead. Despite its precision, the tool can sometimes impact on locations other than their intended places, causing undesirable and unintended mutations. CRISPR-Cas9 is yet to be tried on human beings though it has been successfully tried on other living creatures like mice. A section of the scientists have expressed an apprehension that gene editing tool may ultimately end up in wrong hands and may lead to dangerous consequences. There is also an ethical debate on how far one should attempt to meddle with ones own genes, which have evolved over millennia based on Darwinian evolution by natural selection. However, it cannot be denied that CRISPR-Cas9 marks a quantum leap in our pursuit of understanding the mysterious ways in which human genes behave.

Chapter 14

GLOBAL WARMING AND CLIMATE CHANGE

"It is nice to know that the computer understands the problem but, I would like to understand it too"

– Eugene Wigner

14.1 INTRODUCTION

One of the most pressing problems facing humanity today is how to combat extreme weather conditions which of late have started to occur with amazing regularity. These conditions primarily include extreme temperature variations, draught, rise in sea level and floods. It is naïve to attribute them, as some people do, to the occurrence of natural periodic phenomena unconnected with any human activity. Based on the data collected over several decades, scientists and weather researchers are now convinced that this phenomenon may be directly linked to what is commonly referred to as ***global warming***.

The industrial revolution, which began about 250 years ago, has contributed to a steady build up of gases like, carbon dioxide, Methane and other effluents in our land and water bodies as well as in the atmosphere. As a consequence, the average surface temperature of earth has increased by about 1°C as compared to the average global temperature during the pre-industrial era. This apparently appears to be a small rise in temperature, but what is significant is that half of this temperature rise occurred during the past 40 years. There are abundant examples in nature where very small changes in system initial conditions have led to huge changes in system response. Our global weather phenomenon belongs to one such category.

14.2 WHAT IS GLOBAL WARMING?

We attempt to explain global warming by first recapitulating certain preliminary concepts. Our planet earth is enveloped in a comparatively thin sheet of air known as the atmosphere. It is the gravitational attraction of the earth which keeps the air envelope in position preventing it from escaping into outer space. The major constituents of air are Nitrogen 78%, Oxygen 21% and gases such as carbon dioxide, Nitrous oxide adding up to 1%. Our atmosphere acts as a shield, warding off meteorites and deadly ultra violet rays reaching the earth's surface. As we move up to higher and higher altitudes, air becomes thinner and correspondingly the temperature also steadily decreases. For the purpose of analysis, we may roughly divide the atmosphere into several layers. The layer nearest the earth ranging from 6 Km to 15 Km is known as the *troposphere*. Directly above it is the *stratosphere* which ranges from 15 Km to 50 Km. Above the stratosphere, we have the ionosphere whose properties are of significance while dealing with signal transmission reaching us from outer space.

Our main source of energy comes from the sun whose temperature at its outer periphery is approximately 6000°C. At this temperature the radiation that penetrates earth's atmosphere is mostly confined to the visible range. We are fortunate that this is so, because otherwise our eyes would be incapable of seeing most of the sun's radiation falling on earth's surface. The ultraviolet rays emanating from the sun are effectively blocked by a layer of ozone, which exists in the stratosphere at a height of about 20 Km above the earth. Plants use sunlight to break down carbon dioxide in the atmosphere to carbon and oxygen by a process known as photo-synthesis. The plants use the resulting carbon to build its tissue structure and releases oxygen into the atmosphere. In turn, the animals and human beings breathe in the oxygen and breathe out the carbon dioxide. This ecological cycle is extremely important from the point of view of our very existence on earth. The quantum of carbon dioxide converted by the planet via photosynthesis depends entirely on the forest cover existing on earth. The balance of carbon dioxide escapes into the troposphere where it hangs like a pall, which partially prevents the infrared radiation emanating from the earth and escaping into outer space. Depending on the carbon content in the atmosphere, the average temperature over the earth's surface gradually increases. This phenomenon results in what is known as ***Green House Warming***. It is not the carbon dioxide alone which contributes to green house warming. Methane, Nitrous oxide and moisture content in the atmosphere are also responsible for preventing infra red radiation escaping from the earth's surface into outer space. In the next section we shall examine in more detail the impact of green house effect on our environment in general and the climate in particular.

14.3 DELETERIOUS CONSEQUENCES OF GREEN HOUSE GAS EMISSION

Even with a small 1°C rise in temperature over the pre-industrial level, the impact of Green House Gas (GHG) emissions is now being increasingly felt all over the planet. It manifests itself in many ways. On the one hand, it results in the melting of ice in the Arctic and Antarctic regions, leading to increase in water level in our oceans. This is further accentuated by the warming of the ocean waters and the consequent expansion which results in a gradual rise in ocean levels. This rise leads to the occurrence of frequent floods inundating thickly populated low lying areas. Another unexpected development relates to the "bleaching" of the corals inhabiting the ***Great Barrier Reef***. The algae which reside in these corals are extremely sensitive to the warming sea temperatures. As a result they die *en masse* exhibiting a characteristic white color. The barrier reefs represent the world's largest eco-system with billions of small creatures and marine organisms living on it and global warming will do irreparable damage to this fragile system.

The other manifestations of green house warming relate to frequent drought conditions, unusual rise in average temperatures in certain parts of the world and devastating hurricanes. In this context a graphic description of global warming and its effects are documented in the well known book "*An Inconvenient Truth*" authored by the former US Vice President Al Gore. Can violent hurricanes, floods and droughts be attributed to green house gas effect? Judged by the mounting evidence before us, the answer appears to be an emphatic yes.

14.4 THE OZONE DEPLETION PROBLEM

As already mentioned, the ozone layer in earth's stratosphere, effectively protects us from the lethal ultra violet radiation emanating from the sun. However, in recent times, it was found that that the ozone layer was getting rapidly depleted by chlorofluorocarbons (CFC) emanating from the earth. These chemicals, which are mostly used as refrigerants and in aerosol sprays, have a devastating effect on the ozone layer resulting in what is popularly known as an "ozone hole" right above the Antarctic region. A depletion of ozone will result in the ultra violet rays penetrating this layer and reaching the earth's surface, causing serious ailments like skin cancer among the public exposed to such radiation. Realizing this danger, an international committee which met in Montreal in 1987, banned the use of CFC in refrigerants and aerosol sprays. Known as the Montreal protocol, this banning has had a salutary effect and the ozone layer is now gradually getting replenished. Incidentally, adherence to the ***Montreal protocol*** marks a rare milestone in international cooperation, which has resulted in common good for countries concerned.

While the presence of ozone in the stratosphere is absolutely essential in shielding us from the harmful effects of ultra violet radiation, its presence nearer home in the troposphere poses new health risks, especially among the elderly and those who suffer from respiratory problems.

14.5 STRATEGIES TO CONTROL GLOBAL WARMING

There appears to be a direct co-relation between the quantity of GHG emitted and the global temperature rise. Thus the more the carbon emission, the more will be the temperature rise. It is important to realize that once the carbon content of the atmosphere is increased, it will remain there undiminished for generations to come. In other words, the process is irreversible. There is a tendency among at least a few, belonging to the present generation including scientists, to take things lightly and dismiss the adverse effects of increased carbon content in the atmosphere as nothing to do with human activity. However, it is our bounden duty to keep our planet safe and livable for the coming generations. A brief numerical calculation is given below, which help to illustrate some of the issues raised.

After taking into account past emissions, humanity is left with a carbon credit of approximately 1000 Giga tonnes, if the temperature rise is to be kept within the threshold limit of 1.5°C. With the current carbon emission rate of approximately 40 Giga tonnes per year; this carbon credit will be exhausted in just twenty five years time from now. Thereafter, to maintain the temperature rise at 1.5°C, there should be zero emissions from the earth. This appears to be impractical, as implementing a reduction in carbon emissions is a gradual process and cannot be achieved instantaneously.

The first serious attempt to control carbon emissions at an international level was made in Kyoto, Japan in 1997, at a conference attended by 141 nations. A protocol known as the ***Kyoto protocol*** was signed, which enjoined the participating countries to set limits on their carbon emissions. The cut envisaged was to limit carbon emissions to 5.2% below the 1990 emission levels, and this was to be achieved by 2008-2012. It was noted that 4% of the world population was responsible for 25% of carbon gas emissions, placing an unjust burden on less developed countries to control their emissions, especially when they were not originally responsible for the present situation. However, the Kyoto protocol remained a mere statement of intent, as it was not legally binding on any nation to seriously implement it.

The next land mark effort in this direction was made in December 2015 in Paris, where the historic ***Paris Agreement*** on climate change was drafted. An ambitious target was set to achieve a temperature rise below two degree C or hopefully 1.5°C. Further, it was agreed to provide funds to developing countries towards taking appropriate measures to achieve this objective. At this conference, 191 participating nations volunteered to submit what is known as their individual *"Intended Nationally Determined Contributions (INDC)"*, which unlike on previous occasions was supposed to be subject to periodic verification by an international monitoring body. It was further agreed, that the Paris Agreement could be ratified only if at least 55 countries, whose total carbon emission contributions exceeded 55%, formally signed the agreement. This happened in November 2016, making the Paris Agreement legally binding on all countries that ratified it. A review meeting to consider to what extent, the signatories of the treaty have taken action to adhere to their nationally determined contributions, was planned for November 2020. As on December 2015, the three countries responsible for major carbon emissions globally, were China (26%), USA (16%) and India (6%). Listed below are a few of the countries who have voluntarily agreed to implement their Nationally Determined Contributions by 2030, with a target of 2°C temperature rise.

1. **China** – Agreed to cut peak carbon dioxide emissions by 2027 and cut overall emissions of green house gases by 60-65% of 2005 levels by 2030.
2. **USA** – Agreed to cut emissions by 17% by 2020 and achieve a target of 26-28% by 2025, both measured against a base year of 2005.

3. **India** – Agreed to reduce emissions per GDP (emission intensity) by 33-35% from its 2005 base line and derive about 40% of its total installed electric power from clean energy sources. Further it agreed to create an additional carbon sink equivalent to 3 billion tonnes of carbon dioxide through additional forest and tree cover.

Unfortunately, after ratifying the agreement in 2016, USA has now decided to withdraw from the Paris Agreement, when it is legally able to do so by November 2020. This is indeed a set back to international effort to reduce carbon emissions as proposed by the target date. Should carbon emissions continue to rise after 2020 or even remain static at that level, the goals set up by the Paris protocol appears to be almost unattainable?

14.6 SOME IMPLEMENTABLE DAMAGE CONTROL MEASURES

The technology driven transition to low carbon emission scenario is bound to be a slow process. We now state some of the damage control measures which are feasible and within the reach of all participating nations.

1. **Energy** – The aim should be to make renewable energy to be at least 30% of the world's total electric energy requirement. In 2015, this figure was only 24%.
2. **Transport** – Adopt stringent measures to reduce carbon emissions from transport vehicles by laying down tight regulations especially in the case of diesel vehicles. Another factor to be considered is to improve fuel efficiency. These factors to be made equally applicable to passenger and freight air craft as well. Increasing the number of electrically driven vehicles by a large proportion, should also receive serious consideration.
3. **Afforestation** – Forests play a significant role in reducing carbon emissions by acting as carbon sinks. These points towards the need for rapidly increasing forest cover globally. By the same token, the cutting down of trees under the guise of promoting urban development should be strongly discouraged.
4. **Industry** – Heavy carbon emissions are associated with industries like, steel, cement and oil. One way to cut down these emissions is to improve the operating efficiency of plants through research and development.
5. **Agriculture** – It is not often realized that agriculture plays a significant role in increasing the carbon emissions. These emissions are in the form of Nitrous oxide from fertilizers, and methane gas from live stock and animal husbandry. Clearly, there is a need for planning a global sustainable agricultural policy.

14.7 CONCLUSION

For a 50 Giga tonnes of global carbon dioxide emission recorded in the year 2013, the following per cent wise break up was arrived at.

1. Electric supply (26%) – as a result of burning coal and gas.
2. Industry (19%) – associated with chemical and metallurgical industry.
3. Land use and forestry (17%) – From deforestation.
4. Agriculture (14%) – mainly from biomass conversion and uneconomic agricultural practices
5. Transport (13%) – from burning of fossil fuels such as diesel and gasoline.
6. Commercial and Residential buildings (8%) - arising out of cooling and heating of buildings using air conditioners and heaters and also captive power generation.
7. Methane production (13%) – because of bio-degradable processes and burning of waste.

At the end of the eighteenth century, carbon dioxide in the atmosphere was estimated at 280 parts per million (ppm). By the beginning of the twenty first century this figure had shot up to a figure of 410 ppm. If this rate of increase continues and no corrective action is taken, the carbon dioxide level is likely to reach 1000 ppm by the end of 2100. This alarming figure will give rise to a global temperature elevation of 6°C. Such a situation will be catastrophic, rendering parts of our globe uninhabitable, raising sea levels, causing all round famine due to drought condition and leading to mass migration of population. To prevent this stark scenario from becoming a reality, all nations have to act concertedly and urgently. Not understanding the gravity of the situation and taking an easy view of things is a sure recipe for disaster.

Chapter 15

Artificial Intelligence – Cutting Edge of Modern Technology

"Politics is for the moment, but an equation is for eternity"
– Albert Einstein

15.1 ARTIFICIAL INTELLIGENCE – AN OVERVIEW

Artificial Intelligence (AI) refers to the ability of a digital computer controlled robot to perform tasks commonly associated with intelligent beings. The term is applied to the task of developing systems endowed with intelligent behavior. By intelligent behavior, we mean the ability to discover meaning, to generalize and to learn from past experience. These qualities are generally attributed to human beings.

The word Artificial Intelligence was coined by John McCarthy, a US computer scientist at a conference in Dartmouth in 1956. Apart from McCarthy, the conference was attended by stalwarts in information and computer science, like Claude Shannon and Marvin Minsky. According to them, AI is the science of making intelligent machines, especially intelligent computer programs. This is sought to be achieved by studying how human brain thinks and learns while solving problems. This essentially forms the basis for developing intelligent soft ware systems.

15.2 CLASSIFICATION OF ARTIFICIAL INTELLIGENCE

AI could be categorized in any number of ways, but the most common classification comes under two broad heads- namely ***strong AI*** and ***weak AI***. The former refers to a system with generalized human cognitive abilities with enough intelligence to tackle any unfamiliar problem posed and thereafter to find a solution. Such a system has not yet been fully developed and some experts are of the opinion that this may not happen in the near future.

In weak AI, the system is designed and trained to do a particular job. It is supposed to be a highly efficient system which replicates and perhaps surpasses human intelligence, while performing a particular task for which it is trained. Examples of such systems abound e.g. health care system, self driving cars and chess playing computers.

Early work in AI was focused on solving fairly abstract problems in mathematics and logic. However, this work did not meet with much success. In later years, promising results were obtained in a different context while dealing with huge amounts of data and drawing inferences and patterns from it. While human beings could perform similar tasks, the huge magnitude and time consuming nature proved daunting and put them beyond their reach. Another direction in which research work was carried out was in simulating the human brain. Inspired by the structure of the neurons of the brain, a new area of study known as *Artificial Neural Networks (ANN)*, soon came into existence. Over the years, Artificial Intelligence which originally emerged as a simple concept six decades back has now undergone a metamorphosis. It has made its presence felt in diverse areas like, computer vision, natural language processing, image processing, speech recognition and robotics, to name a few.

15.3 TEST FOR COMPUTER INTELLIGENCE

Earliest theoretical studies on AI were carried out by the English mathematician and computer pioneer Alan Turing, in the mid fifties of the twentieth century. He is the originator of many ideas in AI, which were later on, reinvented by his successors. One such idea related to training a network of artificial neurons to perform specific tasks. Turing was the first to predict as early as 1945, that computers will one day play superb chess to surpass human expertise in this area. This prediction came true fifty years later in 1997, when a chess playing computer called "Deep Blue" built by IBM, beat the then reigning world champion Gary Kasparov in a six game contest. However, this achievement may largely be attributed to the advances in computer technology rather than to new developments in AI. Indeed, a renowned expert in linguistics Noam Chomsky of MIT was reported to have remarked, "The success of the chess playing computer was as interesting as a bulldozer winning an Olympic weight lifting competition".

To assess the intelligence capabilities of the computer, Alan Turing proposed a test, which is now known as the *Turing Test*. The Turing Test involves three participants namely, a computer, a human interrogator and another human, who simultaneously responds to questions posed to the computer. All the three are isolated in separate rooms and the means of communication with them is through computer key board and display screen. The interrogator is free to ask penetrating questions on a wide range of topics to both the computer and its human counterpart. After posing a series of questions, going by the answers provided by the human and the computer, if the interrogator is unable to distinguish between the human and the computer then the computer is considered to be a thinking intelligent entity.

In the next section, we shall explore in some detail, the role played by Artificial Neural Networks (ANN) as a tool in AI studies. We begin with the structure of the biological neuron and then show how its functions can be simulated by an artificial neuron.

15.4 NEURON CELLS ARCHITECTURE

The human body houses about 200 different types of cells. One such type of cell is the neuron found in the brain and the nervous system. Neurons are responsible for carrying sensations such as pain, pleasure etc, through molecules known as neurotransmitters.

Fig.1 shows the typical structure of a biological neuron. It has a globular body called *soma* with a head that resembles a flower with hundreds of petals. Each of them is called a *dendrite*. At the other end a neuron has got a long stem called *axon*. The axon fans out at its ends into hundreds of terminals. In effect, a neuron resembles a tall palm tree. Its leaves correspond to dendrites and its roots are the axon terminals. The cell body houses the nucleus. It is typically about 50 micrometers in diameter. The axon (occasionally one meter long) resembles a tube with diameters ranging from 0.2 to 20 micrometers. The axon provides the main conducting path for the neuron.

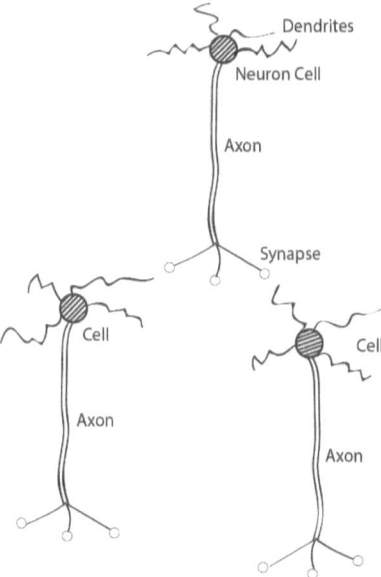

Fig 1: Neuron cell and its connections

Considering a typical neuron (referred to here as the first neuron for identification), its cell body receives pulses from other neurons via numerous dendrites dotted on its surface. Based on the total signal received from other neurons, the first neuron adds on its own response, and together the signal travels along the axon to its end terminals. These terminals end up in what are known as *synaptic sacs*. The synaptic sac of the first neuron is separated from the receptor site of a neighboring neuron namely its dendrite by a microscopic distance. The cell body of the first neuron produces chemical substances called *neurotransmitters* which travel down the axon and are delivered to the synaptic sacs. When the first neuron fires, the neurotransmitter molecules are released from the synaptic sac and attach themselves to the dendrites of the neighboring neurons, where the two fit each other like a glove. The first neuron has several such synaptic sacs. The neighboring neurons have several such receptor sites and this is how signals are propagated throughout the neural network. Every cell is associated with a threshold value of the signal and it will fire only if the instantaneous accumulated value of the signals received from outside, exceeds this value. Considering that a single neuron has about thousand synapses and there are millions of neurons in the human brain, the complexity of these connections is mind boggling. Clearly the Artificial Neural Networks which are supposed to simulate the brain cannot possibly compete with this type of complexity even with the help of the most powerful super computers. Incidentally, while discussing neural transmitters in a different context when the neural pathways get blocked as the result of snack bite, the venom thus injected into our bodies results in total paralysis and ultimate death.

How Artificial Neural Networks attempt to simulate the brain function is the subject matter of the next section.

15.5 CREATING ARTIFICIAL NEURAL NETWORKS (ANN)

The aim in building ANN is to simulate the human brain and its functioning. The ANN comprises of a number of artificial neurons which are linked together or networked to form a structure to represent the brain. The artificial neurons (as opposed to biological neurons) is a simple device which receives signals from various sources, with respective weights (the weights may represent the relative intensity of the signals). They add together and this sum is compared to a pre assigned threshold value of the signal. If the weighted sum exceeds this threshold value, then the neuron fires and this is denoted by symbol 1. On the other hand if the threshold value is more than the sum of the weighted signals, then no firing will take place and the event is denoted by symbol 0. Fig 2, shows a simplified version of the artificial neural network.

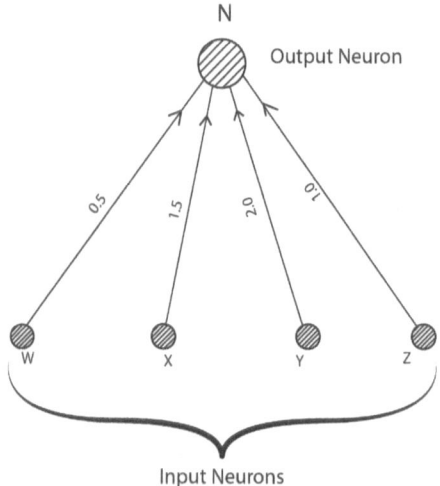

Fig 2: Artificial Neural Network

The simple neural network illustrates the central idea behind the working of ANN. Four out of the five neurons marked; W, X, Y and Z in the figure serve as input terminals, while the fifth neuron N is the output terminal. Each connection which leads to N has a weight associated with it. The total weighted input to N at any instant is arrived at by adding up the weights of all connections leading to N. For example, if W, X, Y and Z fire in quick succession, then according to Fig 2, the weighted inputs into N is 5.0 and this being greater than the pre-assigned threshold value of 4 for N, the neuron will fire.

Training of Neural Networks

To perform certain tasks for which it is intended, an ANN should undergo what is known as a training schedule. To start with a series of input and output pairs (x_1, y_1), (x_2, y_2) .... (x_p, y_p) are specified where x_i is the ith input and y_i the corresponding ith output. In an untrained network an input x_i in the first instance will not give rise to the specified output y_i. The idea is to go on adjusting the weights of the network till the actual output is as close to the specified output as desired. This procedure has to be repeated for every one of the p- input-output pairs and further repeated many times over after one cycle is completed. In most cases this exercise is not done manually but through a computer. The striking feature is that the method is carried out using a computer algorithm. There is no external intervention at all and the correction weights are increased or decreased automatically to achieve the desired objective.

A simple illustrative example will suffice to make things clear. Consider a neuron N, fed by several weighted inputs. The desired output of the neuron is either a 1 when it is firing or 0 when it is not firing. This is achieved through a two step process.

Step-1: For a given input, if the actual output is 0, while the desired output is 1, then increase the several weights involved by small amounts in successive trials till the output is 1 and the neuron starts firing.

Step-2: Alternatively, if the actual output to start with is 1, whereas the desired output is 0, then successively reduce the weights by small amounts and repeat the process till the output registered is 0, i.e. the neuron stops firing.

15.6 MULTILEVEL NEURON NETWORKS

In a more general neural network we can have apart from the input layer and the output layer, several layers of neurons in between. Here again we can draw a distinction between a *feed-forward neural network* and a *back-propagation neural network*. In the feed forward case all information signals flow from input to output in one direction. Fig 3, shows a feed forward neural network with two hidden layers. Input neurons receive signals from the environment and in turn send signals to the neurons in the hidden layer.

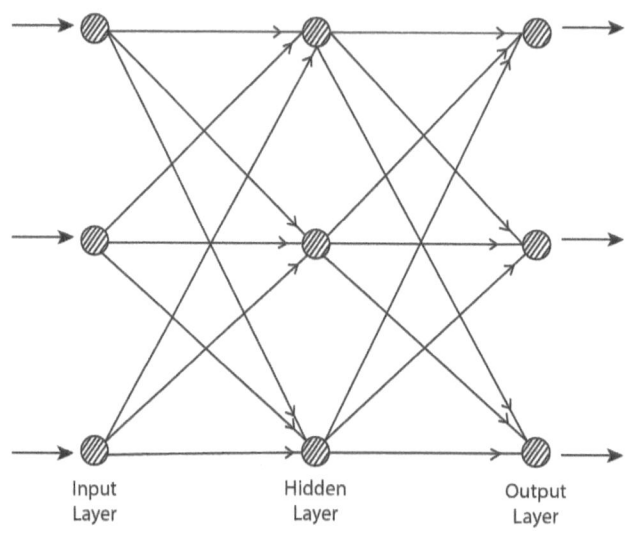

Fig 3: A feed forward multilayer neural network

Whether any particular neuron, sends a signal to fire or not, depends on the combined signal strength of signals received from preceding layers. A multi layer network is the basis for back propagation network. As the name implies, there is a backward pass of error signal flow for each internal node which is then used to calculate the weight gradients for that node. The network learns by alternately propagating backward, the errors as and when they occur. Back propagation can deal with various types of data and has the ability to model a complex decision system.

15.7 MAJOR ARTIFICIAL INTELLIGENCE AREAS

Major areas in AI comprise of Expert systems, Natural language processing, Speech recognition, Robotics, Neural computing etc,. The various tools employed in this connection are Evolutionary computing, Fuzzy logic and Neural Networks. We shall briefly consider here some of the areas mentioned.

Expert Systems – It uses AI methods to solve problems within a specialized domain that ordinarily requires human expertise. An expert system has two components, namely a *knowledge base* and an *inference engine*. The knowledge base is acquired through human experts via interviews and observations. The knowledge base of a major expert system incorporates thousands of rules. A probability factor is often built in to the formulation of each rule. An expert system may also if required display the sequence of rules through which it arrived at a particular conclusion.

An inference engine interprets and evaluates facts in the knowledge base taking into account the large volume of information provided to it. Expert systems have the ability to learn from experience as more data are acquired. However, expert systems only remain as aids rather than replacement for human experts.

Natural Language Processing – (NLP) refers to an AI method of communicating with an intelligent system, using natural language such as English. Processing of natural language is absolutely essential if one wants to communicate with an intelligent system like robots to perform specific tasks. NLP is a method which enables the computers to analyze, understand and derive meaning from human language. This is extremely difficult as the human language is rarely precise and to understand the language means not only the meaning of individual words, but also how the underlying concepts are linked together. This ambiguity of human language is what makes natural language processing a difficult problem for computers to master.

Speech Recognition – It is a technology that allows the computer to identify and understand words or commands spoken by a person through a microphone or telephone. With all the progress made in this area, the best recognition software application still cannot recognize speech with 100% accuracy. Software that understands speech enables us to have conversations with the computer. This facility eliminates to some extent physical devices such as keyboard and mouse pointing devices. Further, this device helps the visually handicapped person to interact with the computer to a limited extent.

Robotics – It refers to the use of computer controlled robots to perform manual tasks. Robots are now routinely used to perform tasks that are dangerous, if performed by humans such as, defusing bombs, repairing gas leaks or even unmanned reconnaissance flights. They are also used to perform monotonous jobs in assembly lines in a factory. Today, robotics is a rapidly growing field which has found its application even in simplifying household chores. Continuing research has enabled us to make smarter and more capable robots. A popular development in this connection is the deployment of driverless cars. The robot drives more efficiently and reacts faster than humans, have 360° vision and they do not get distracted, sleepy or intoxicated. Another area where robots find extensive application is in the medical field. Robot- assisted surgeries are now commonplace allowing for a steadier hand in surgery and more accurate interpretations. Water- based robots can now perform underwater surveillance under hazardous conditions. The latest application of AI is in the design of the so called drones, which can sneak into enemy territory, take photographs of strategic locations and details and safely return to their bases.

15.8 EVOLUTIONARY COMPUTATION AND GENETIC ALGORITHMS

Evolutionary computing belongs to a family of algorithms for global optimization inspired by biological evolution and natural selection. To start with an initial set of candidate solutions is generated by the computer and they are iteratively updated.

Each new generation is produced by removing less desired solutions and introducing small random changes. In biological terms a *population* of solutions is subjected to *natural selection* and *mutation*. As a result the population will gradually evolve and ultimately only the fittest will survive. Evolutionary computational techniques can produce highly optimized solutions to a wide range of problems. Closely allied to this field are the so called *genetic algorithms*. It is a process in which a set of solutions to the problem called the "population" is iterated. Each, iteration brings a new generation of the population. The fitness of every individual in the population is evaluated against the objective function embedded in the problem. The more fit individuals are then selected from the latest population and their *genomes* modified to form a new generation. The new generation of candidate solutions is then used in the next iteration of the algorithm. The algorithm terminates when after a number of iterations, a satisfactory fitness level has been reached for the population.

Fuzzy Logic – Fuzzy logic is a form of many valued logic in which the value of the variable may range anywhere between 0 and 1. It is employed to handle concepts of partial truth, where the truth values may range between completely true or completely false. This is in sharp contrast to *Boolean logic* where the truth value can only be either 0 or 1. *Fuzzy set theory* was introduced in 1965 by Lofti Zadeh as an extension of the classical notion of a set. Fuzzy logic has been applied in many fields ranging from control theory to artificial intelligence. Its significance as far as artificial intelligence is concerned stems from the fact that ANN may be used to implement fuzzy systems.

15.9 FUTURE OF ARTIFICIAL INTELLIGENCE

At the rate at which AI is developing, it appears certain that in the next few decades most of the tasks which we now perform will disappear, because it will be cheaper, quicker and far more efficient to have these performed by robots and artificial intelligence. The AI systems are bound to have a profound effect on what we do and how to do it in the future.

A major limitation of AI is that, in the ultimate analysis it can learn only from the data available and if this happens to be wrong, this will be reflected in the end results. A second drawback is that AI systems are trained to perform specific tasks only. A system trained for manipulating driverless cars cannot by any stretch of imagination perform medical diagnosis. In this respect at least for the present AI is inferior to human intelligence. A third indirect consequence is that it will render a majority of our work force jobless and this is bound to have major repercussions on the functioning of our society.

Much as the benefits due to advances in AI are, there is an alternate point of view which has to be noted. According to some, AI poses the greatest risk for our civilization. An eminent physicist like Stephen Hawking believes that AI will ultimately take over the world and end the human race. This opinion voiced by a person of his stature, who paradoxically has benefited personally from the advances in AI, is highly significant. Though he is not personally against the implementation of AI, the danger he sees ahead deserves due consideration. According to him, AI in the future will become smart and powerful enough to surpass human beings in every conceivable respect. Ultimately, it will take off on its own and redesign itself at an ever increasing rate. These criticisms notwithstanding, AI now as well as in the future will continue to remain at the cutting edge of modern technology.

Chapter 16

EXPLOSIVE GROWTH OF INFORMATION TECHNOLOGY AND COMMUNICATION SYSTEMS

"Where is the wisdom we have lost in knowledge?
Where is the knowledge we have lost in information?"

– *T.S Elliot*

16.1 THE RISE AND RISE OF INFORMATION TECHNOLOGY

We live in an era where we are bombarded with information from all directions. The written word, the radio, the TV and recently the internet are constantly pouring out information and one is hard put to digest even a fraction of what is being dished out. Out of all this morass has emerged a new technology now commonly referred to as *Information Technology (IT)*. It is primarily a by product of 20th century scientific advances. In IT, we can clearly see the convergence of electronics, data processing and telecommunications.

This convergence has two aspects. First, the emergence of the Internet and the World Wide Web have virtually demolished time and distance. Second, the development of computers with their extremely fast acting response and burgeoning memory have enabled us to tightly pack information and deliver them in the form of sounds and images. It is truly amazing that by a mere press of a key or a button, the contents of an entire library are placed at our disposal, immaterial of whether we are functioning at home or at a for away place cut off from all modern amenities. Information technology has superseded the cumbersome mechanical and electro-mechanical processes and relieved human beings of much of the burden of routine work which they are otherwise accustomed to. The explosive growth of IT in recent times may mainly be attributed to developments in semiconductor technology and the miniaturization brought about by the development of Integrated chips (IC) whose capacity for performing computation has almost doubled every two years since 1965 (recall Moore's law). What is most significant is that, the gestation periods of the new technologies have proved to be remarkably short when compared to earlier times. To put these developments in proper perspective, we propose to carry out a rapid fire survey of how IT developed in earlier days and its exponential growth in recent times.

16.2 HISTORICAL PERSPECTIVE

Today's communication technology is synonymous with radio, telephone and the internet. However, these developments mark the culmination of a long drawn out process - a process which lasted many centuries. First, it was the development of language among human beings, solely for communication purpose and this happened centuries before the birth of Christ. Next came the development of written language involving symbols and characters which occurred around 2000 B.C. The development of the postal system came next. The mode of communication in earlier times employed messengers on foot or on horseback to deliver the message. The method was prevalent in Egypt and China as early as 1000 B.C. However, this procedure proved inadequate, especially for the wide spread dissemination of information. The necessity of having books was thus recognized as early as third millennium B.C, in the form of clay tablets and parchment scrolls, by the Mesopotamians and the Egyptians. However, the major development in this direction took place some time between 1440 AD and 1450 AD, when Johannes Gutenberg invented the printing press. This was followed by the publication of the Gutenberg bible in 1454. Subsequent to the invention of writing, dissemination of knowledge came largely through the publication of books. Some of the science literature which continues to inspire human thought even to this day were published during the 17th and 19th century such as Isaac Newton's *"Principia Mathematica"* (1687) and Charles Darwin's *"Origin of Species"* (1859).

The science of communication experienced a major thrust with the invention of telegraph by Samuel Morse in 1837. Before this invention, information could only move as fast as the physical means deployed in transporting it. It took weeks and often months for communication to reach from place to place. It seemed as though with the invention of the telegraph, all of a sudden space and time have dramatically shrunk. However, Morse's invention required that the natural language had first to be transcribed into a code before it could be transmitted. This code, well known as the *Morse code* is composed of a succession of dots and dashes. It served as the backbone for telegraphic communication. This system however, had its own drawbacks. The message had first to be taken to the telegraph office where it had to be coded and then transmitted via telegraph lines to reach its destination.

Some of the shortcomings inherent in transmitting messages via telegraph system were overcome with the invention of the telephone, by Alexander Graham Bell in 1875. The use of the telephone made person to person communication possible, doing away with the coding system and also the visits to the nearest post office to receive or deliver a message. However, the mode of transmission employed, which was basically analog in character had to depend on telegraph wires criss- crossing the country for the message to reach its destination. This invention also had its own limitations and was slow in catching up. For example, the results of the British expedition's successful observations of the solar eclipse (May 1919) in Africa (which confirmed Einstein's theory of relativity), was conveyed to Einstein only after a lapse of 3 months in September 1919, even though telegraphic communication was available at that time.

The dream of wireless communication became a reality with Gugliemo Marconi's famous experiment in 1894, when he successfully transmitted a signal from USA to England across the Atlantic ocean. This was the first practical demonstration of radio transmission over a long distance. At that time there was no understanding of the reflective properties of the ionosphere which made this discovery all the more creditable. The wireless propagation of signals soon led to developments in radio and television.

16.3 OPTICAL FIBER COMMUNICATIONS

As wireless media expanded, so did the capacity to transmit vast amount of data through this medium. What started as an original idea of telephone conversation using a pair of copper wires evolved into a system of thousands of conversations which could be carried out by optical pulses via a glass fiber. This marked the birth of fiber optic cables which proved to be a vastly superior and highly efficient medium for communicating information from point to point. An optical fiber cable is similar to a coaxial electrical cable but contains instead of copper wire, one or more optical fibers to do the job. It consists of a core and cladding with different refractive indices, so chosen as to enable total multiple internal reflections to take place. The signal propagates along the cable through a series of such reflections. The fiber optic communication system in its simplest form consists of a transmitter and a receiver with the fiber optic cable acting as a transmission line linking them. The source is optical and consists of a *Laser diode* and an appropriate driver's circuit. The information signal to be transmitted modulates the source signal. The signal propagates through the fiber and is then demodulated at the receiving end. The reason why ordinary light cannot be used for such purposes instead of the laser beam is because ordinary light is composed of many wave lengths traveling in different directions and is highly diffused and incoherent. Modulation and demodulation for signal transmission cannot be done because of the inherent randomness of the ordinary light wave. This problem was elegantly solved with the discovery of the Laser. It essentially consists of beams of only one wave length which reinforce each other and move only in one direction. It is therefore highly coherent and ideally suited to be used as a carrier wave for modulation and demodulation purposes. Fortunately, right at the moment when the need arose for the availability of such a beam, Laser was invented in 1958 by Arthur Schawlow and Charles Townes. Two years later, Theodore Maimen built the first practical laser optical cables to transfer data via laser at the speed of light in glass. The round trip delay time for 1000 Km of optical cable is around 11 milliseconds. Like all transmission systems, fiber optic transmission is also subject to attenuation losses. A typical figure for attenuation loss in long distance transmission, is about 0.25 dB/km at 1550 nanometer wave length. The attenuation is taken care of by the installation of amplifiers at designated intervals all along the line. The vast majority of all long distance transmission system now uses fiber optic cable. Submarine fiber optic cables are used to connect land based stations separated by vast oceans. To summarize, fiber optic cables have a much greater band width than cables using metallic conductors and further the power loss is small. This allows for longer transmission distances.

16.4 THE ELECTRONICS REVOLUTION

In 1947, the transistor was invented at Bell labs USA by William Shockley, John Bardeen and Walter Brattain. This laid the foundation for the electronics revolution which gained momentum in subsequent years. The transistor was first used as a direct replacement for the vacuum tube having a much smaller size and requiring much less power to operate. The next major development was the fabrication of a chip with several transistors on a substrate. This heralded the birth of the integrated circuit. The use of the integrated chip in microprocessors, personal computers and mobile phones followed in quick succession. The microchip development was predicted as early as 1965 by Intel's cofounder Gordon Moore. Moore's law predicts that the number of transistors that can be embedded on a microchip doubled every two years, thanks to advances in photo lithography and other fabrication tools. This trend has continued up till now. However, this cannot go on indefinitely as inter atomic distances and other physical laws will come in the way of further miniaturization.

The electronics revolution has fostered the development of mini sized computers. The rise of computer networks and the digitization of information media followed in quick succession. The striking feature of these developments is that the incubation period i.e. the period between the discovery of an idea and its practical application has dramatically reduced in recent years. Today's World Wide Web (www) can been cited as a classic example to illustrate this point. Compared to the earlier belief that major technological developments will require at least 25 years for its wide spread availability, the World Wide Web has arrived in double quick time. However, it must be borne in mind that the web required wide spread availability of computers, which in turn depended on the advances in microprocessor technology. These developments could ultimately be traced to the discovery of the transistor.

16.5 THE INTERNET AND THE WORLD WIDE WEB

The *internet* is a massive network which links together millions of computers globally thus forming a huge network in which any computer can communicate with any other computer as long as both are connected to the internet. Information travels along the internet in a variety of languages known as *protocols*. Each internet computer is independent of the rest of the computers in the net. The World Wide Web is a way of accessing information using the medium of the internet. It is an information sharing model built on top of the internet. The web uses *HTTP protocol,* one of the languages used in the internet to transmit data. The web utilizes *browsers* such as *internet explorer* or *fire fox* to access web documents called web pages. They are linked to each other via the so called *hyper link.* The web documents may contain graphics, such as text and video. Web browsers make it easy to access the World Wide Web often known as the web.

The web represents just one of the ways that information can be disseminated over the internet. The internet (not the web) is used for e- mail transmission. The web therefore occupies only a portion of the internet albeit a large portion but the two terms are not synonymous. The global address of documents is given by the **URL** (*Uniform Resource Locator*) which makes it easy to visit addresses or documents in the World Wide Web. When we search for information through Google for example, the search result will display the URL of resources that relate to the search query.

The most important application in early days of the internet turned out to be simple e-mail exchanged between researchers. Historically, the World Wide Web came into existence from a physics laboratory at the CERN where it was pioneered by Tim Berners Lee in 1989. Collaboration among scientists today is largely carried out through the internet and the web. Useful as the web pages are to disseminate information, it must be noted that the information available is often not peer reviewed and therefore must be treated with some caution.

16.6 WEB SEARCH ENGINES

A web search engine is a software system that is designed to search for information in the World Wide Web. The search is generally presented as a set of results referred to as a search engine result pages. Typically, when a user enters a query into a search engine, the query is limited to a few key words. The usefulness of a search engine depends on the relevance of the result it provides. While there are millions of web pages that include a particular word or phrase, some pages are more relevant, popular or authoritarian than others. How a decision is made about which of the pages are the best match for a query, depends on the search engine used. Most search engines are commercial ventures, supported by advertising revenue. One of the popular search engines is the *Google.* The others include Yahoo, Bing, AOL.com, etc. It is no exaggeration to say that we have reached a stage where we depend on search engines for almost any human activity.

16.7 INTERNET: SOCIAL MEDIA

Social media in the internet is user generated. It includes comments, digital photos, videos and data generated through on line interactions.

User Generated Content (UGC) plays a role in a wide range of applications including news, entertainment, research and even gossip. It illustrates a classical example of the democratization of the network content at the most basic level. People from all walks of life are now able to post text, digital photos and videos on line free of cost without being subjected to the tyranny of censorship. Some of the prominent social media network sites include, Face Book, What's App, Twitter, Instagram, Snap chat, Skype, Linked-in etc,. The user generated content of these sites is immensely popular, especially among the youth. However, these social sites have attracted criticism from skeptics who have questioned its apparent lack of fairness, low quality of presentation and intrusion into the privacy of individuals.

16.8 PACKET SWITCHING

Packet switching describes a network transmission process in which data is broken up into relatively small packets and routed through the network based on the destination address contained in each packet. Breaking down data into packets allows the same data path to be shared among many users in the network. The packets carry the data in the protocol that internet uses namely TCP/IP (Transmission Control Protocol/Internet Protocol). This is the language that the computer uses to access the internet. Packets find their way to their destination with the help of routers, which uses their knowledge about the network at any instant to control the movement of each and every packet along the most efficient route. Different packets arrive at the destination along different routes in the network at different times. Since, every packet contains the destination address and the packet number; they are reassembled in the correct sequence at the destination. A typical packet may contain perhaps 1000 to 1500 bytes. This method of transmitting information is more efficient and results in a better utilization of the network. In the early days of packet switching it was mainly used to transfer real time audio and video communication. Since the voice and video traffic are routed through the same channel used for data transmission, there is no need to have separate computer and phone networks. This results in considerable cost saving.

16.9 CONCLUSION

With the rise of high speed internet connections and web enabled cell phones, it is now easier than ever before to access information in the internet. As people turn to internet for news and information, traditional media sources like newspapers, magazines, encyclopedias etc, are slowly becoming obsolete. The World Wide Web has had the greatest impact on society compared to any communication medium since possibly the invention of the printing press. The postal system, telegraph and telephones are no different from each other except for the fact that the last two are much faster. Further, the three systems mentioned are *one to one*, involving only the sender and the receiver. On the other hand, radio and television result in *one to many* communications, as a smaller number of channels cater to a larger number of people. The internet and web belong to an entirely different category. Their topology is neither one to one nor one to many, but *many to many*. This is a quantum jump from the earlier modes of communication. It has opened up the flood gates to millions of people who can now share information in textual, graphic and multi-media formats at all times and in all locations.

The infrastructure provided by the internet for research and collaboration in science is now rich and full of promise. Who would have imagined two decades back that we could have access to an entire library by merely pressing a key of a lap top computer or hand held device, in the privacy of our home. Remote sharing of facilities provided by costly instruments and laboratory set up are now possible. Video conferencing over the internet now facilitates discussions among scientists who are located in different parts of the globe. One can visualize a future world where all transactions will be on-line with minimum of human intervention. Thanks to information technology and progress made in the development of communication systems, exciting times lie ahead of us.

Chapter 17

NUMBER THEORY – UNFOLDING A FASCINATING TAPESTRY

"We have heard much about the poetry of mathematics but very little of it has yet been sung"

– Henry David Thoreau

17.1 INTRODUCTION

According to the eminent mathematician and philosopher Bertrand Russell

"Mathematics rightly viewed possesses not only truth but supreme beauty- a beauty cold and austere like that of a sculpture, without appeal to any part of our weaker nature, without the gorgeous trappings of paintings or music, yet sublimely pure and capable of a stern perfection, as only the greatest art can show".

This quote admirably sums up all that mathematics stands for. There is no better example to demonstrate the above aspects than Number Theory – rightly labeled by Gauss as the *"Queen of Mathematics"*. The intimidating transcendental beauty of Number Theory has attracted during the past hundreds of years, great mathematicians like Gauss, Riemann, Euler and Ramanujan, to name a few.

Numbers offer a fascinating field of study. The underlying concepts are simple but some of the conjectures posed are deep and profound. It has baffled the best of mathematicians for centuries. Numbers are purely a creation of the human mind, but its impact has been felt by every branch of human knowledge. It is in this spirit that a chapter exclusively devoted to number theory has been included in this book, which primarily explores modern scientific thought.

17.2 RECAPITULATION OF BASIC TERMINOLOGY

In mathematics, a real number is a value that represents some quantity. These numbers can be indicated on an infinitely long line stretching from minus infinity to plus infinity. The adjective real was introduced in the 17th century by Rene Descartes to distinguish between real and imaginary roots of polynomials. Real numbers include integers -2, +5 etc., fractions $\frac{7}{11}, \frac{3}{8}$ etc...; irrational numbers $\sqrt{2}, \sqrt{5}$ etc...; and transcendental numbers such as e, π etc....

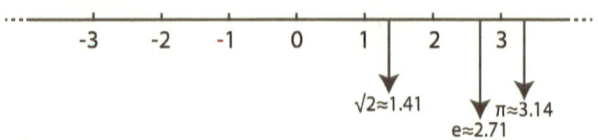

Fig 1: Representation of real numbers on the number line

Real numbers can be broadly categorized into two classes – rational and irrational.

Rational Numbers are those which can be expressed as a quotient of two integers - p and q. The decimal expansion of a rational number always either terminates after a finite number of digits or repeats the finite sequence of digits over and over again. Examples are: 7/5 = 1.4; 1/3 = 0.3333....etc. Hence any terminating or repeating decimal number come under the category of rational numbers.

Irrational Numbers are those real numbers which are not rational. Examples are: $\sqrt{2}$, $\sqrt{3}$, e, π etc. The decimal expansion of an irrational number continues indefinitely on and on without any pattern repeating itself.

A certain sub-class of irrational numbers are grouped under the head *transcendental numbers*. They cannot be obtained as the roots of a polynomial equation with integers or equivalently rational coefficients.

Examples are: π = 3.1415........., e = 2.7182.......

$\sqrt{2}$ for example is an irrational number that can be derived as one of the positive roots of the polynomial equation: $x^2 - 2 = 0$. Similarly, the golden ratio φ which has a value φ = 1.6180......, can be derived as the positive root of the polynomial equation $x^2 - x - 1 = 0$.

Prime number is an integer which cannot be formed by multiplication of two or more smaller integers. A prime number is divisible only by itself and unity. Number 1 is not a prime number by definition.

An integer greater than 1 which is not a prime is called a *composite number*. The first few prime numbers are: 2, 3, 5, 7, 11, 13, 17, 19, 23, 29 etc...... There are infinitely many primes located along the real number line as shown by Euclid as early as 300 BC.

17.3 DISTRIBUTION OF PRIME NUMBERS

Prime Numbers are the building blocks of the number system. Any composite number can be considered as the product of two or more primes. Thus the number 34866 = 2 x 3 x 3 x 13 x 149, where all the indicated factors are primes. The distribution of prime numbers have intrigued mathematicians for centuries. There appears to be no discernible pattern in their distribution along the real number line. The gap between consecutive primes is one of the problems that have attracted the attention of number theorists. Table-1 gives the distribution of primes in the range of integers 1 to 1000.

Table 1: Distribution of Prime Numbers for Integers ranging from 1 to 1000

2	3	5	7	11	13	17	19	23	29
31	37	41	43	47	53	59	61	67	71
73	79	83	89	97	101	103	107	109	113
127	131	137	139	149	151	157	163	167	173
179	181	191	193	197	199	211	223	227	281
283	293	307	311	313	317	331	337	347	349
353	359	367	373	379	383	389	397	401	409
419	421	431	433	439	443	449	457	461	463
467	479	487	491	499	503	509	521	523	541
547	557	563	569	571	577	587	593	599	601
607	613	617	619	631	641	643	647	653	659
661	673	677	683	691	701	709	719	727	733
739	743	751	757	761	769	773	787	797	809
811	821	823	827	829	839	853	857	859	863
877	881	883	887	907	911	919	929	937	941
947	953	967	971	977	983	991	997	-	-

A close examination of Table-1 shows that the gap between the first few consecutive prime numbers seems to grow steadily from 1 to 6 before tapering down to 2. This pattern more or less continues till the gap jumps to 14 between consecutive primes 113 to 127. For several primes occuring thereafter, the gaps are mostly small with occasional spikes going up to 20. However when we reach the 217th prime given by the number 1327, there is a sudden jump in the gap by 34 from 1327 to 1361. This jump goes on increasing erratically as we proceed along the number line.

Several questions arise at this juncture:

1. Are there any gaps between consecutive prime numbers with a specific value (e.g with gap 2) which goes on repeating indefinitely as we proceed towards infinity?
2. Given any integer n, can we ascertain by any known formula, the number of primes up to and including n?
3. Given an integer n is it possible to assign a lower bound to the number of primes up to and including n?
4. Is there any universal formula that can generate all prime numbers starting from the beginning up to infinity?

The patterns that the distribution of prime numbers exhibit in the number scale (if at all there is a pattern), have still not been fully understood. Table-1 serves to illustrate some of the vagaries in the distribution of prime numbers, especially in the beginning stages. Some pattern seems to emerge as the integers are increased indefinitely. This will be explained in the subsequent section with the help of what is known as the Prime number theorem.

17.4 THE PRIME NUMBER THEOREM

In number theory, the prime number theorem describes the asymptotic distribution of prime numbers among the positive integers. It formalizes the intuitive idea that primes become less common as they become larger by quantifying the rate at which it occurs. As we shall see in the sequel, some of these questions have partial answers and some others as of now, have no answers.

This fundamental theorem of prime numbers was proved independently by Jacques and Charles Poussin in 1896. It gives an approximation of the number of primes less than or equal to any given positive real number x. In this connection the terms involved are defined as follows:

$\pi(x)$ = The actual number of primes, including x for a given positive real number x.

$\log x$ = The natural logarithm of x

The *prime number theorem* (PNT) states that for very large values of x,

$\pi(x)$ may be approximated by $x/\log x$ i.e. $\pi(x) \sim x/\log x$

Another way of stating the same result is,

$\pi(x)/x \sim 1/\log x$

Table-2 gives the proportion of the actual number of primes to the given number x and approximation of this ratio as given by $1/\log x$. It can be seen from the Table that as x becomes very large; this approximation comes very close to the actual value.

Table 2: Comparison between predicted and actual proportion of primes

Number "n"	$\pi(n)$	$\pi(n)/n$ (Actual)	$1/\log n$ (Predicted)
10^2	25	0.2500	0.0217
10^4	1229	0.1229	0.1086
10^6	78,498	0.0785	0.0724
10^8	5,761,455	0.0570	0.0543
10^{10}	455,052,511	0.0455	0.0434
10^{12}	37,607,912,018	0.0377	0.0362
10^{14}	3,204,941,750,802	0.0320	0.0310

Table-2 is revealing in many respects. First as the numbers increase, the density of occurrence of primes actually decreases. Thus when the number under consideration is 1000, there are 168 primes within that number. Out of 1000 only a very small fraction (168/1000 = 0.168) is occupied by primes. As seen from the table the average distance between primes goes on increasing, which in reality implies that primes become scarcer for large values of n. A second inference from the table is that as n becomes large, 1/log n is a good approximation of the ratio π (n)/n.

17.5 THE TWIN PRIME GAP CONJECTURE

The conjecture relates to the magnitude of gaps between successive primes. As already mentioned, when the integers increase, the average gap between successive primes also increases. However, even when considering very large numbers, it is noticed that there occasionally occurs primes with a twin gap. A famous conjecture in number theory asserts that this twin gap will continue to occur occasionally for ever and for ever. Even though this conjecture is yet to be proved, a related highly significant result was proved by US number theorist Yitang Zhang. He showed that there exists a number N in the less than 70 million number range, which will give rise to infinitely many prime gaps with gap N. This result was further refined by James Maynard and Terence Tao who between them independently showed that the bound of 70 million could be drastically reduced to a mere number 246. This result still does not show that the gap N predicted is actually 2 or some other number within 246. Another interesting result, which was recently independently proved by James Maynard and a few others relates to the largest gap between consecutive primes up to and including a number x under consideration. They have shown that a lower bound exists for this gap, in the sense that the largest gap attained should be greater than this lower bound. In other words as the number x under consideration increases, the largest gap among consecutive primes cannot be less than the lower bound arrived at by Maynard and others for a given x. However, the greatest lower bound in this connection is yet to be proved.

17.6 THE MERSENNE NUMBER

From time to time number theorists have come up with formulae which generate a set of primes, though they do not cover all the primes in the real number field. One such result pertains to the so called *Mersenne Numbers*, named after the 17th century French mathematician Marin Mersenne. A Mersenne prime is a prime number resulting from the formula $M_n = 2^n - 1$, for some integer n. The exponent n which yields Mersenne numbers are 2,3,5,7,13.......and the corresponding primes are 3,7,31,127,8191...It can be proved that the exponent n should be a prime in order to yield prime numbers. But not all prime exponents give rise to prime numbers. For example, the prime number n = 11, gives rise to $2^{11} - 1 = 2047 = 23 \times 89$. This clearly shows that n = 11 which is a prime, does not yield a Mersenne prime number. The largest prime number known so far (as of Dec 2017) is generated by the Mersenne number is [$2^{77,232,197} - 1$]. This number has 23,249,425 digits – a million digits larger than the previous Mersenne prime generated This number also happens to be the 50th Mersenne prime generated so far.

17.7 FERMAT'S LAST THEOREM

Fermat's last theorem is a conjecture which states that no three positive integers a, b, and c satisfies the equation:

$a^n + b^n = c^n$, for any integer values of n greater than 2. This theorem was first conjectured by the French mathematician Pierre Fermat in 1637. He wrote the above equation in the margin of a book and stated that he had the proof of this conjecture, but it was too large to fit in the margins of the book. For over 358 years several eminent mathematicians struggled to obtain a proof for Fermat's last theorem. Some number theorists did arrive at proofs for specific values of n, but to prove it for all integer values of n, appeared formidable. Finally, in 1995, Andrew Wiles a British mathematician published a proof which once and for all showed that Fermat's conjecture was indeed correct. The proof is of great importance as it uses many techniques from algebraic geometry and number theory and has wide ramifications. The proof published by Andrew Wiles was described as a "stunning advance" in the citation, given to him on the occasion of his receiving the prestigious Abel's Prize in 2016.

17.8 APPLIED NUMBER THEORY

Number theory was once upon a time considered the purest of pure mathematics, so much so that the eminent number theorist the late Leonard Dickson remarked, "*Thank God that Number theory is unsullied by any application*!". Had he been alive today, he would be in for a surprise as the situation has vastly changed in recent times. Prime numbers and composite numbers now play an important role in cryptography and coding systems. With the advent of the internet, high volume of confidential information (credit card and bank account numbers) and large sums of money are electronically being transferred around the world every day. It is very important to keep these transactions secret, to prevent hackers from stealing sensitive and confidential information. Cryptography allows users (government, military, businesses and individuals) to maintain privacy and confidentiality in their communications and business transactions. To create a security system, first the letters and characters of a message are numerically coded. Once the message is translated into numbers, an algorithm multiplies these numbers by a certain prime number (which may be a 100 digit number) and by a composite number (a product of two prime numbers randomly selected). This composite number which may be quite long (100 digits or longer) will be available in the encoding algorithm of the sender and decoding algorithm of the receiver. When the messages are transmitted from the sender to the receiver, some numbers are made public, but the primes that make up the composite number are kept secret. These are only known to the person who receives the messages. For any one who is eves dropping on the information transmitted, without the prior knowledge of the prime numbers associated with the encoding and decoding programmes, it is impossible to decode the message in a reasonable time. Number theory has got a lot of applications relating to network security. One of the widely used cryptosystems for secured data transmission involves what is known as the RSA (Rivest – Shamir – Adelman) algorithm. In this system the encryption key is public and is different from the decryption key which is kept secret. In RSA this asymmetry as in the previous case, is based on the practical difficulty in the factorization of the product of two large prime numbers.

Chapter 18

LIMITS OF SCIENCE

"Science becomes dangerous only when it imagines that it has reached its goal"
— George Bernard Shaw

18.1 THE METHODOLOGY OF SCIENCE

What is the underlying spirit of scientific enquiry? This is succinctly described by Carl Sagan an ardent advocate of science and a fierce opponent of pseudo science and mysticism. Science he writes is different from any other human enterprise,

> *".... in its passion for framing testable hypothesis, in its search for definite experiments that can confirm or deny ideas, and in the vigour of substantial debate and in its willingness to abandon ideas that have been found wanting"*

According to Karl Popper, – one of the 20th century's greatest philosophers of science, science progresses by a series of *"conjectures and refutations"*. Popper believed that a theory can only be disproved and can never be proved. A theory becomes scientific by exposing itself to the possibility of being incorrect. Liability of empirical disproof is the defining character of science. He believed that any theory that was not *"falsifiable"* i.e. capable of being tested and proven incorrect should be dismissed as unscientific. The falsification principle is the corner stone of modern scientific method. However, some contemporary scientists believe that this principle needs to be revisited, as in the case of theories like String Theory and Multiverse, which come up against the possibility of being testable, at least for now. A physical theory or a picture is generally a model of a mathematical nature that prescribes a set of rules that connect the model to the observations. There is no question of absolute reality associated with the model and any reality is model-dependent. If there are two models that both agree with the observations, we cannot say one is more real than the other.

Some of the criteria of a good scientific model are, it should be elegant and it should contain few arbitrary or adjustable constants. Apart from matching existing observations, it should also be capable of making detailed predictions of the future that can disprove or falsify the model, if the observations are not borne out.

18.2 AN ALTERNATIVE VIEW POINT

As opposed to the scientific approach, there are a few who believe that the universe is the handiwork of an intelligent creator. In the 18th century a British clergyman, William Paley likened the wonders of the universe to the working of a complex watch. As in the case of a watch with many parts, designed to perform specific tasks all working in unison, the universe relentlessly ticks on. Just as a watch requires a watchmaker Paley identified the creator of the universe as its watchmaker.

This argument was challenged by Richard Dawkins, an evolutionary biologist from Oxford in his book "The Blind Watchmaker". Though written in a biological context, the argument he advances is relevant in its application to other areas as well. Dawkins in his book vividly explains the rationale behind Darwin's theory of Natural Selection. According to him, the blind watchmaker is nature itself, gradually seeking order in a process which is automatic, blind and yet essentially non-random. A group of science skeptics, who call themselves as *creationists*, dispute this view and believe in what is commonly referred to as *intelligent design*. They reject Darwin's theory of evolution and natural selection and assert that all evolution takes place at the discretion of an intelligent designer. This is a rehash of the old "God - of - the Gap" argument according to which any phenomenon which is not satisfactorily explained by science at the moment is attributed to the will of God. However, it looks as though the God-gap is continuously closing all the time.

18.3 THE ANTHROPIC PRINCIPLE

An argument which is some times put forward even by scientists, (though it is more akin to intelligent design), to justify certain phenomenon, relate to what is known as *anthropic principle*. It has long been observed that the various constants associated with nature such as the gravitational constant, Planck's constant, etc, are so fine tuned that even a small departure from their normal values will turn out to be disastrous for existence of life on earth. However, life on earth being a reality, their argument goes that the constants in question have been deliberately adjusted by the maker of the universe with the sole purpose of making life on earth possible. But this argument apparently fails if the concept of multiverse is accepted, in which case there may exist thousands of other worlds where human beings may not exist and where the constants and even the laws of physics are free to vary widely.

18.4 SCIENCE VERSUS RELIGION

The arguments in favor of science adduced so far do not mean that science has no limitations. According to Arthur Eddington the well known British astrophysicist, theories accepted by science could only address those features of the universe which are quantifiable. However, there are elements of human experience which cannot be explained by either science or mathematics. Religion, aesthetics, compassion, love etc, come under this category. According to Eddington, religious experience is no less real than physical experience and should solely be judged by the individuals concerned. Based on this philosophy, Eddington rejected any claim by individuals to prove or disprove religion by scientific methods. Religion and science relate to two different realities of experience that can neither validate nor support each other. Both involve continual searches for knowledge, the former in the spiritual domain and the latter in the physical world. Searching and not finding constitute the basis for both religion and science. The above opinion, coming from a leading scientist did find a favorable echo among many contemporary scholars of his era.

18.5 BELIEVERS AND NON-BELIEVERS

It is fair to state that among the scientific community there are both believers and non-believers. The non-believers assert that life in general and human life in particular is totally irrelevant against the backdrop of a vast and meaningless universe. Therefore to assume that some higher power is guiding our day to day activities, appears to be irrelevant. As opposed to this, the believers present a different view point. They see a grand designer at work behind the cosmic evolution, who they identify as God or creator of the universe. Scientists exhibit a variety of dispositions and as the renowned biologist Peter Medawar put it, "*Among scientists are collectors, classifiers and compulsive tidiers up; many are detectives by temperament and many are explorers; some are artistes and some are artisans. There are poet scientists, philosopher scientists and even a few mystics*".

We recall below some of the beliefs entertained by leading scientists over an extended period of time. Starting with Galileo, the father of modern scientific approach, we learn that he was a deeply religious person. But this did not prevent him from expounding scientific theories, which ran counter to the then prevailing beliefs of the Catholic Church. The church condemned him for heresy for insisting that the earth revolved around the sun, instead of the other way round. However, it took more than 400 years for the Pope to acknowledge this error, accept Gilileo's theories and rehabilitate him, in 1992. Focusing our attention next on Isaac Newton, one of the greatest scientists the world has ever known, his biographer's report that he was highly superstitious and exhibited strong mystic leanings, came as a great surprise. When Newton was asked why all the planets orbiting the sun moved in the same plane, he had no explanation and attributed this phenomenon to "*divine providence*". We now know that the planet formation arose out of a spinning nebula of gas and dust surrounding the proto sun and this was responsible for the phenomenon referred to.

Another view point was, that attributed to the famous 18th century mathematician Laplace, who wrote the classic book titled "Celestial Mechanics". It is reported that Napoleon after scanning through the book asked Laplace, why he has not made even a single mention of God anywhere in this book, Laplace is supposed to have replied, "*your honour, I had no need for God hypothesis*". When Napoleon repeated this conversation to Lagrange, another great contemporary mathematician, he is reported to have remarked "*But it is a fine hypothesis - It explains so many things*".

Turning our attention next to a cross-section of 20th century scientists, they exhibited a wide spectrum of opinions and beliefs, ranging from atheism, agnosticism and surrender to the will of the divine. In this respect, Einstein was extremely articulate as revealed by some of his statements. Commenting on the probability aspect of quantum mechanics, he is supposed to have remarked, "*God does not play dice with the world*". His approach to nature is summed up in the lines – "*Subtle is the Lord, but malicious he is not*". When asked to explain his statement, he remarked that, "Nature hides her secret because of her essential loftiness, but not by means of ruse". Another famous quote from Einstein, runs as follows – "*Science without religion is lame and religion without science is blind*".

Apart from Einstein, many of the leading scientists of the 20th century had their own opinions about matters other than science. For example, Schrödinger was influenced by Eastern religious thought and Niels Bhor believed in the notion that living organisms in some ways transcended the laws of physics.

Most of the opinions quoted here happen to come from physicists. By and large they are against dogmas and preconceived notions associated with religion. For example the physicist Steven Weinberg says in his 1977 book titled, "The First Three Minutes"– "*The more the universe seems comprehensible, the more it also seems pointless*". He went on a step further to assert that, "*Anything that we scientists can do to weaken the hold of religion should be done and may in the end be our greatest contribution to civilization*". Sentiments of a similar nature were voiced by another celebrated physicist, Richard Feynman when he said, "*The greatest accumulation of understanding of how the physical world behaves, only convinces one that this behavior has a kind of meaninglessness about it*". Last but not the least, the opinion of the eminent cosmologist Stephen Hawking is worth quoting. In an interview given to the press in 2007, Hawking said that he was "*not religious in the normal sense*". He went on to add, "*I believe the universe is governed by the laws of science. The laws may have been decreed by God, but God does not intervene to break the laws*".

The above quotations seem to belie the concept of a universe driven by a cosmic creator, bent upon moving it towards certain preordained goals.

18.6 CONCLUSION

In the final analysis, it looks as though among scientists, the atheists and agnostics outnumber the faithful. Admittedly, there are two ways of knowing the world- through religion and through science. Religion cannot contribute to the scientists understanding of the world. Likewise, science cannot explain the meaning and philosophy of life. A single minded pursuit of science does not make a person an atheist. There are several scientists who are deeply religious but at the same time believe in evolution. In short, science and religion constitute two realms of experience. Scientific enquiry and practice of religion are two basic impulses of the human mind and what makes a good scientist could also make a good religious minded person.

Glossary

Note: The items listed here are arranged chapter wise in no particular order. Many of those listed here have already been explained in their respective chapters, but, some are explained in more detail. This glossary is primarily intended to quickly recapitulate the essence of the technical jargon found in the main text.

RELATING TO CHAPTERS 2 TO 9

Big Bang Model:
This is the currently accepted model of the universe according to which time and space emerged from an extremely hot dense and compact region about 13.8 billion years ago.

Big Crunch:
The name given to one possible scenario of the universe where all space and, matter collapse to form what is known as a singularity.

Atom:
The smallest component of an element comprising of a positively charged nucleus, surrounded by negatively charged electrons. The nucleus is composed of protons which are positively charged and neutrons with zero charge. The number of protons in the nucleus, uniquely determines which chemical element, the atom belongs to. For example, if an atom has 79 protons in its nucleus, then it refers to an atom of gold. On the whole a neutral atom has zero charge.

Isotope:
It refers to an element, with a fixed number of protons associated with that element, but having different number of neutrons in its nucleus. For example, the element Hydrogen has three isotopes corresponding to zero, one and two neutrons, but all of them contain just one proton. The isotopes in this case are hydrogen, deuterium and tritium.

Galaxy: A collection of stars, gas and dust held together by gravity and separated from neighboring galaxies. Galaxies range in size from around a million stars to billions of stars. Our solar system belongs to a galaxy called the Milky Way which contains about a billion stars. The galaxies come in either spiral or elliptical shapes.

Hubble's Constant (H_0):

It is a measurable parameter of the universe describing its rate of expansion. The current value of H_0 appears to be 73.2 Km per sec per mega parsec, thereby meaning that a galaxy 1 mega parsec away from us will be receding at the rate of 73.2 Km per second. The Hubble's constant arises from the definition of Hubble's law.

Hubble's Law:

This experimentally determined law states that the velocity at which galaxies are receding is proportional to the distance between them. This is given by the relationship, $v = H_0 d$, (where H_0 is the Hubble's constant, v is the velocity of recession of the galaxy from observer and d the distance between the observer and the galaxy).

Astronomical Unit (AU):

The mean distance between the earth and the sun as the former orbits the sun.

1 AU = 1.5×10^{11} meters

Par Sec (Pc):

It is equal to the distance at which 1 AU subtends an angle of 1 arc sec along the vertical axis ($1/3600^{th}$ of a degree)

For example, Proxima Centauri – our nearest neighboring star is at a distance of 1.3 pc from us. Andromeda galaxy our nearest galactic neighbor is located at a distance of 0.71 Mega pc from us. Sagittarius A*- a super massive black hole at the center of the Milky Way is about 30,000 light years away from us.

Light Year:

The distance traveled by light at a speed of 3×10^8 meters/sec in one calendar year.

1 par sec = 3.26 light years.

Doppler Effect:

It is the change in wave length of sound or electromagnetic wave emitted by a moving source.

A familiar example is the change in pitch of a siren from high to low as an ambulance passes by a stationary observer.

Red Shift:

The apparent radiation emitted by an object which is moving away from an observer, as a result of Doppler effect.

Inflation:
Refers to a phase of extremely rapid expansion of the universe hardly 10^{-35} seconds after the big bang and which lasted for about 10^{-30} seconds.

Even though inflationary universe is a hypothesis, it explains several verifiable features of the universe.

Cosmological Principle:
It is a principle that states that no location of the universe is preferred over any other location on a large scale. This implies that:

1. The universe appears to be the same in all directions (isotropic)
2. No matter where the observer is located in the universe, it appears to be the same (homogenous)

Cosmological Constant:
An extra parameter incorporated by Einstein into his equation on general relativity when the equation clearly predicted either an expanding or shrinking universe. As he found this result to be incompatible with his belief of a static universe, he arbitrarily introduced the cosmological constant, to make the universe appear static. He later on disowned the cosmological constant calling it his "greatest mistake".

Copernican Model:
The sun centered model of the universe (also known as the heliocentric model) proposed by Nicholas Copernicus in the sixteenth century.

White Dwarf:
It is all that is left of a star (like our sun), which has died and ejected its outer layers. The whiteness is attributed to the extremely hot exposed core of the star which may shine brightly for a while, before it gradually cools off and fades away.

Supernova:
It is an extremely powerful and energetic stellar explosion which may shine with the brightness of an entire galaxy. Supernovae are of several types. Type 1a supernova happens when in a binary system (a white dwarf and an ordinary star orbit each other), stellar material is transformed into a white dwarf.

In the type-II supernova, the explosion occurs due to death and collapse of massive stars several times the mass of the sun.

The Chandrasekhar Limit:
Named after its original proposer Subramanyan Chandrasekhar, the limit states that a white dwarf can at the most support 1.4 solar masses. Beyond this mass, the white dwarf star collapses as in the case of type-1a supernova explosion.

Cosmic Microwave Background Radiation (CMBR):
It is the after glow of the intense energy radiation which started radiating outwards about 380 million years after the big bang. While the radiation commenced at a temperature of about 3000K, today due to the expansion of the universe it has cooled down to 2.7 K, which is what we measure on earth. The original wave length of the emitted radiation is in the infra red region, but today the wave length has stretched into the micro wave region, with its peak occuring at about 2 mm.

Cepheid Variable:
This refers to a class of variable stars whose brightness varies over a period usually between 1 and 100 days. The period of variation is directly linked to the stars' average luminosity which can be calculated. By comparing the stars luminosity to its apparent brightness as seen from earth, its distance can be determined.

Kuiper Belt Objects (KBO):
The Kuiper belt stretches in space from between 30 AU to 50 AU from the sun far beyond the orbit of Neptune. KBO refers to the vast number of small objects orbiting the sun at about 40 AU from the sun.

Oort Cloud:
An imaginary sphere of radius of about 2/3 light years, enclosing the entire solar system. It is the origin of *long period* comets with highly elongated elliptical orbits in which they come close to the sun and then get back to the very edge of the solar system. Their orbital period may be as large as thousands of years.

Nebula:
Vast clouds of gas and dust occupying the interstellar space. Galaxies, stars and planets are formed out of these gas clouds under the influence of gravitational force.

Neutron Star:
It is formed during the last moment of a massive star's life. As the massive star dies, the electrons and protons are quashed together inside the core under the influence of intense gravitational force to form a super dense ball of neutrons, hardly 20 Km across, known as the neutron star.

Pulsar:
It is a spinning neutron star that continuously fires jets of radiation from its magnetic poles. Because the stars magnetic axis is often offset from its axis of rotation, the jets appear to sweep around at tremendous speeds, as in the case of light emanating from a light house.

Quasar:
Also known as the quasi stellar radio objects, they are associated with enormous super massive black holes located at the centre of galaxies devouring matter at a furious pace. As the spinning gas and dust collect around the black hole, it loses vast amounts of gravitational energy. This causes huge streams of highly energetic radiation to blast away from the black hole, usually making the quasar many times brighter than its host galaxy.

Circum Stellar Disc/Proto Planetary Disc:
It is the disc of gas and dust surrounding a young proto star, and rotates about it. As the rotation increases, collisions occur among bodies within the disc, sticking them together. This ultimately results in planet formation.

Event Horizon:
Marks the outer boundary of a black hole. Any particle crossing this boundary, not even light can escape the gravitational pull of the black hole.

Gamma Ray Burst (GRB):
They represent some of the most powerful explosive phenomenon occuring in the universe. They are caused either when two neutron stars or black holes collide or possibly when a giant star explodes.

Meteoroid:
It is a piece of planetary debris perhaps no bigger than a small pebble or rock. They are left over during the formation of the solar system.

Meteor:
Also known as the shooting star, it is seen during a brief moment when a meteoroid passes through the earth's atmosphere. During this process, it gets heated up and then burns quite harmlessly as they touch the earth.

Meteorite:
They are pieces of natural space debris that frequently land on earth. They are usually made up of metals like iron and nickel as well as rock. They are the best ways the planetary

geologists can study the surface of a planet or any other solar system body without actually visiting them. Studying the meteorite can tell us about the early solar system formation 4.5 billion years ago.

Electromagnetic Radiation:
It covers the entire spectrum of radiation, starting from radio waves through visible light, X-rays and gamma rays.

Observational Tools Used for Cosmology Parameter Measurements:
1. Gamma Ray Telescope.
2. X-ray telescope- To measure high energy photons belonging to X-rays. They cannot travel a long distance through the atmosphere without alteration and therefore has to be observed from beyond the atmosphere or in outer space.
3. Ultraviolet wave length measurements. These waves are associated with wave lengths ranging from 10 to 230 nano meters and are normally absorbed by the thin ozone layer in the stratosphere before they reach the earth.
4. Visible rays – visible ray astronomy extends from 400 to 700 nano meters. They are measured by optical telescopes which are either ground based or satellite borne. The latter contributes to better clarity as the observations are not clouded by atmospheric distortions.
5. Infrared astronomy- The rays measured are above 700 nm and are associated with cooler objects.
6. Sub-millimeter astronomy- observations at sub-millimeter wave lengths are often made by satellite mounted telescopes orbiting at low altitudes.
7. Radio waves – They are measured by radio telescopes an instrument designed to detect radio waves. Their wave lengths are in the region of a few millimeters and above. This includes micro waves also. As the atmosphere is transparent to radio waves, radio telescopes are generally ground based. Radio telescopes are nothing but highly sensitive radio receivers and consist of an antenna or a dish of appropriate size
8. Microwaves – Microwave space telescopes are mostly used to measure cosmic background radiation (CBR). The COBE (Cosmic Background Explorer) satellite launched in 1989 was intended for the sole purpose of making such measurements.
9. Gravitational waves – These waves are due to the ripples caused in space time, as a result of colliding neutron stars and black holes. They are extremely weak and are propagated as waves emanating outward from their sources at the speed of light. Gravitational wave astronomy is a branch of observational astronomy

which uses gravitational waves to collect data about the supernova and the formation of early universe shortly after the big bang. The Laser Interferometer Gravitational Wave Observatory (LIGO) searches for distortions in space-time that would indicate the passage of gravitational waves.
10. Neutrino observatory – Unlike other observatories which look towards the sky, the neutrino observatory is buried deep underground, searching for particles known as neutrinos. These are nearly massless particles and interact weakly with matter. They offer a unique probe into some of the most violent processes occuring in the universe involving neutron stars and black holes. The neutrino detector in the India based Neutrino observatory (INO) is proposed to be located in a tunnel almost 1300 meters below the ground under Bodi West Hills, about 100 Km from Madurai, India.

Nucleus:
The central part of an atom consisting of protons and neutrons in the nucleus held together by what is known as the *strong force*.

Nucleon:
It's a generic name given to denote either protons or neutrons in the nucleus.

Neutron:
Is an uncharged particle very similar to the proton. It is composed of 3 quarks (2 down and 1 up).

Neutrino:
Is an extremely light particle that is affected only by *weak force* and gravity.

Proton:
It is a positively charged particle, very similar to a neutron. It is made up of 3 quarks (2 up and 1 down).

Electron:
It is a particle with negative electric charge that orbits the nucleus of an atom

Quark:
It is a charged particle that feels the *strong force*. Quarks come in 6 flavors, namely up, down, strange, charm, bottom, top. Each flavor has 3 colors- red, green and blue.

Fermion:
It is a particle whose associated *spin* is one half of a whole number. For example, electrons, protons, neutrons etc, belong to the family of fermions.

Boson:
It is particle whose associated *spin* is a whole number. For example, photons, gluons, gravitons etc, belongs to the family of bosons.

Antiparticle:
For every particle, there is a competing antiparticle which has the same mass but carry equal and opposite charge. When a particle collides with an antiparticle, both are annihilated, leaving only energy behind. For example, positron is a positively charged antiparticle of the electron.

Dark matter:
Refers to matter in galaxies and possibly between clusters of galaxies that cannot be observed directly but can be detected by their gravitational influence. About 90% of matter of the universe is in the form of dark matter.

Dark energy:
It is a postulated form of energy that could possibly account for the recent observation that the universe is expanding at an accelerated pace. It accounts for about 70% of the mass-energy content of the universe.

Four Fundamental Forces of Nature:
1. Strong force – The strongest of all the four fundamental forces with the shortest range. It holds quarks together to form protons and neutrons and further holds them together to form the atomic nucleus.
2. Electromagnetic force – The force that acts between particles with electric charges of similar or opposite sign. It has infinite range.
3. Weak force – The second weakest of the four fundamental forces with a very short range. It affects all matter particles, but not force carrying particles like photons, gluons etc, known by the generic term *bosons*
4. Gravitational force – The weakest of the four fundamental forces, but with infinite range

Boson:
It is a particle which belongs to the category of fundamental force carriers and has integral *spin*.

Photon:
Is a quantum of light energy, the smallest packet of the electromagnetic field. It is a force carrier for electromagnetic field and has zero mass, zero charge and unit spin.

Gluon:
The force carrier in the case of strong interactions – It has zero mass, zero charge and unit spin.

W & Z Bosons:
These are the force carriers in the case of weak interactions. There are 3 different types of Bosons-

1. W^+ boson – with unit positive charge, mass of 86 GeV and unit spin.
2. W^- boson – with unit negative charge, mass of 86 GeV and unit spin.
3. Z boson with zero charge, mass of 97 GeV and unit spin.

Fusion:
It is the process by which two small atomic nuclei join together to make a single large nucleus, while releasing energy as a consequence. The enormous energy produced by the sun and the stars at extremely high temperatures is attributed to the fusion process. For example, hydrogen nuclei fuse together via a multiple step process to form helium inside the core of the sun.

Fission:
It is the process by which a large atomic nucleus is broken apart to produce smaller nuclei and generally releasing energy as a result. Most of the nuclear power reactors functioning are based on this principle.

Radioactive decay is a fission process that occurs spontaneously as in the case of uranium.

Nucleosynthesis:
The formation of elements via nuclear fusion particularly in stars and supernova explosions. The big bang nucleosynthesis occurred hardly a minute after the big bang, when hydrogen ions and neutrons combined to form helium ions, with the resulting matter a mix of 75% hydrogen and 25% helium.

Plasma:
A high temperature state of matter in which atoms are separated from their orbiting electrons to form ions

Vacuum Energy:
Energy that is supposed to be present even in apparently empty space. It has the strange property that unlike in the case of mass energy, vacuum energy's presence will cause the expansion of the universe to speed up.

RELATING TO CHAPTERS 10 & 11

Special Theory of Relativity:
Based on Einstein's premise that the speed of light is the same for all observers, regardless of their own motion. It also implies that the perception of time and space depends on the observer. The theory is called "special" because it does not deal with objects that are accelerating or experiencing gravity.

General Theory of Relativity:
Unlike in the case of special theory of relativity, the general theory proposed by Einstein deals with accelerated reference frames. In particular, it describes gravity as a curvature in space-time.

Time Dilation:
It is a feature of the special theory of relativity predicting that a clock in a moving object with respect to a stationary observer, ticks more slowly than without such motion. This is also true in the presence of a strong magnetic field.

Gravitational Lensing:
This relates to an effect where light from a distant object, for e.g. a quasar is distorted (or lensed) as it encounters a large mass such as a galaxy located nearer to an observer

Lorentz Contraction:
The shortening of moving objects along the direction of motion as predicted by the special theory of relativity.

Mass:
The quantity of matter in a body, leading to its inertia or resistance to acceleration in free space.

Planck's Quantum Principle:
The idea that electromagnetic waves for example light waves can be emitted or absorbed only in discrete quanta.

Planck's Length:
About 10^{-35} centimeters.

Planck's Time:
About 10^{-43} seconds, the time it takes light to travel the distance of Planck's length.

Quantum Mechanics:
Physical laws that govern the realm of the very small, such as atoms, protons, neutrons and electrons.

Quantum Gravity:
A theory which merges general relativity with quantum mechanics.

Uncertainty Principle:
A principle formulated by Heisenberg which states that one can never determine simultaneously both the exact position and the exact velocity of a particle. The more accurately one knows about the position of a particular particle, the less accurate will be our knowledge about its velocity and vice-versa. This can be rephrased by stating that, the product of uncertainties in position and velocity, must be greater than $h/2\pi$, where h is the Planck's constant.

Exclusion Principle:
The idea that two identical spin ½ particles cannot have (within the limits of the uncertainty principle) both the same position and same velocity, is known as Pauli's exclusion principle. This does not hold for *bosons* which are integral spin particles.

Wave/Particle Duality:
A concept in quantum mechanics which says that there is no difference between a particle and the wave associated with it. Depending on the situation, a particle may behave as a wave or vice versa.

Schrödinger Equation:
The equation governing the evolution of the wave function in quantum theory.

Schwarzchild Radius:
It refers to the radius R of a black hole's event horizon where $R = 2GM/c^2$. Here G stands for gravitational constant, M for the mass of the black hole concentrated at the centre of a sphere of radius R and c the speed of light. Thus the radius R is proportional to the mass of the black hole.

Black Body Radiation:
Black body is a theoretical concept where the radiation energy flowing into a body is the same as that flowing out, thus keeping the body in thermodynamic equilibrium. Depending on the energy flow, in and out, this equilibrium is attained at different absolute temperatures. The frequency spectrum of the incoming and outgoing radiation

can be widely different even though the total energy per unit time may be identical in both cases.

Wien's Displacement Law:
When the black body radiation spectrum is examined, for a given absolute temperature T of black body radiation, the maximum energy density occurs at a certain wave length λmax. As the temperature T is increased, λmax decreases.

This inverse proportionality relationship is given by λmax T = constant,

Thus for a higher and higher absolute temperatures, the peak value of energy density radiation occurs at smaller and smaller wave lengths.

RELATING TO CHAPTERS 12 & 13
DNA – (Deoxyribo Nuclic acid):
It is a very long molecule that contains instructions for living things to build and maintain themselves. All organisms have their own unique strands of DNA in each cell forming a very long code called genome. DNA has got a double helix structure. This was discovered by two biologists Crick and Watson in 1953. This may be likened to a very long twisted ladder with two back bones connected at intervals by the rungs of the ladder. Each back bone contains chemical building blocks namely A, C, G & T not in a particular order, located at regular intervals. They are known as nucleotides. A stands for Adenine, C for cytosine, G for guanine, T for Thymine. The nucleotides or *bases* on either side of the ladder are linked by the rungs of the ladder according to what is known as base pairing principle. Thus base A links only with base T and base C links only with base G. The complex code ACGT arranged in various combinations along the backbone of the ladder is different for different individuals except in the case of identical twins. A single strand of DNA is 3 meters long and yet it fits into each cell's nucleus which has only a volume of 2 to 3 cubic microns (1 micron = 10^{-6} meters).

Genome:
The complex DNA sequence of an organism is known as the genome.

Genes:
Certain structures of an organism's genome, performs certain functions. These sections are known as genes. Certain genes are responsible for the production of proteins essential for the functioning of all living organisms. However, all genes do not encode proteins. Examples of proteins are insulin, hemoglobin etc. Human genomes were completely mapped in 2003 as a result of the epoch making Human Genome Project. Surprisingly human genome contains only about 20,000 protein coding genes as against 50,000 genes in rice, which has an extremely simple structure compared to humans.

Chromosomes:

DNA of every organism is tightly packed into "X" shaped bundles known as chromosomes. There are 46 chromosomes in every human being out of which 23 chromosomes are inherited from the father and 23 from the mother, which accounts for the 23 pairs of chromosomes. The first 22 pairs of chromosomes are more or less of the same size but the 23rd pair is quite different. This pair, known as the sex chromosomes, are also identified as X and Y chromosome. In a female the 23rd pair consists of two X chromosomes whereas in a male they consist of an X and Y chromosome. During fertilization when the sperm fuses with an egg if the child's father contributes an X chromosome, then the child will be a female with XX chromosome pair. On the other hand if the father contributes a Y chromosome, then the child will be a male with XY chromosome pair. The sex chromosome X houses about 1000 genes, whereas the sex chromosome Y is comparatively small housing about 27 genes.

The unique genome of every human being carries 3 billion base pairs from each parent.

RNA (Ribo Nucleic Acid):

RNA is a ribonucleic acid that helps the synthesis of proteins in our body. RNA resembles DNA, the only difference being that it has a single strand unlike DNA which has two strands. RNA is also referred to as an enzyme since it helps in the process of chemical reactions in the body. It has identical bases namely, Adenine, Guanine and Cytosine as that of DNA, except that Thymine is replaced by Uracil.

RNA has two major functions. First, it assists the DNA and serves as a messenger between DNA and the Ribosome. Secondly, it helps the ribosome to choose the right amino acid which is required in building proteins in the body. There are two types of RNA in the human body, namely t-RNA (transfer RNA) and m-RNA (messenger RNA).

m-RNA is responsible for carrying the genetic material to the ribosomes.

t-RNA is responsible in choosing the correct protein or amino acid required by the body.

Ribosomes:

Ribosome is a complex molecular factory found inside the living cells that make proteins from amino acids. Every cell needs ribosomes to manufacture proteins. It binds to the m-RNA and reads the information contained in the m-RNA.

Telomere:

Chromosomes carry our genetic material. If one considers them as shoe laces, telomeres are little protective tips at its end. They are repetitive short DNA sequences sheathed in special proteins. During ones life time, telomeres wears down continuously and cannot protect chromosomes properly. This sets up physiological changes in the body contributing to our aging process. The shortening of telomeres due to age is a natural process but this process can be slowed down by adopting a proper life style, such as stress management, exercising and good dietary habits.

Stem Cells:
A stem cell is a cell with the unique ability to develop into certain specialized cell types of the body. Our body is made up of many different types of cells. Most cells are designed to perform particular functions, for example red blood cells carry oxygen around our body. Some of these cells are unable to replicate themselves. Stem cells provide new cells for the body as it grows and also replaces specialized cells that are damaged or lost. They have unique properties:

1. They can divide over and over again to produce new cells to replace damaged cells.
2. As they divide they can change into other types of cells that make up the body.

Pluripotent Cells:
Pluripotent cells are cells which can give rise to or *differentiate* to all cell types that make up the human body. In this respect, embryonic stem cells are considered to be pluripotent. As against pluripotent cells, there are multipotent cells which can only develop into certain type of cells. For example, adult stem cells are considered to be multipotent.

Induced Pluripotent Stem Cells (IPS cells):
IPS cells are a type of pluripotent stem cells artificially derived from a non-pluripotent stem cell- by external "inducement" or forcing. The ability to induce adult stem cells into a pluripotent state was initially pioneered in 2006, by molecular biologists Yamanaka and John Gurdon. These induced cells exhibit similar traits to those of embryonic stem cells and can replace embryonic stem cells in certain cases.

Epigenetics:
It is the study of non-inherited changes in the gene function caused by that which does not involve change in the DNA sequence. The epigenetic changes may last through cell divisions, for the duration of the cells life.

Retrovirus:
It is a type of virus that uses RNA as its genetic material. After infecting a cell, a retrovirus uses an enzyme called reverse transcriptase to convert RNA into DNA. The retrovirus then integrates its viral DNA into the DNA of the host cell, which in turn allows the retrovirus to replicate. For example, the HIV virus that causes AIDS is a retrovirus.

CRISPR Technology:
It is a simple but powerful tool for editing genomes. It allows researchers to alter DNA sequences and modify gene function. Its many potential applications include correcting genetic defects.

RELATING TO CHAPTERS 14, 15, 16 & 17
Photosynthesis:
Photosynthesis is a process used by plants and other organisms to convert light energy into chemical energy. The resulting chemical energy is stored in carbohydrate molecules. In most cases oxygen is also released as a byproduct. In photosynthesis, occuring in plants, carbon dioxide plus water in the presence of sunlight is transformed into carbohydrates and oxygen. The carbohydrates produced are stored in or used by plants. Photosynthesis is largely responsible for producing and maintaining oxygen content of the earth's atmosphere.

Green House Effect:
The sun's radiation reaching our planet is mostly in the visible range. This radiation passes through the earth's atmosphere to reach us. However, the radiation emitted by the earth is predominantly infrared and this is partially blocked by the earth's atmosphere. This results in the warming of earth's surface to a temperature above what it would be without earth's atmosphere. This phenomenon is known as Green House effect.

Green House Gases:
Green House gas refers to a gas in the atmosphere that absorbs and emits radiant energy within the infrared range. This property is the fundamental cause for the green house effect. The primary green house gases in the earth's atmosphere are water vapor, carbon dioxide, methane, nitrous oxide and ozone. Without the green house gases, the earth's average surface temperature would be -18°C instead of the present average of 15°C.

Ozone Layer:
The ozone layer or ozone shield is a region of earth's stratosphere that absorbs most of the sun's ultraviolet (uv) radiation hitting the earth. It is a thin layer which consists of high concentration of ozone (O_3). The ozone layer is mainly found in the lower portion of the stratosphere in a 20 to 30 Km wide range. In the absence of ozone layer, the UV rays will directly impact on earth causing potential damage to exposed life forms on earth.

Chlorofluorocarbons (CFC'S):
CFC's commonly known by the brand name Freon, was widely used as refrigerants and propellants in aerosol applications. It is one of the primary causes for the depletion of the

ozone layer in the upper earth's atmosphere. Because of this harmful effect, it has been phased out under the Montreal protocol and replaced by other products such as hydro fluorocarbons (HFC's), which are less harmful.

El Nino:
El Nino is an abnormal weather pattern caused by the warming of the Pacific Ocean near the equator, off the coast of South America. El Nino occurs every 3 to 5 years but can even come as frequently as every 2 years or rarely every 7 years. It often causes severe changes in weather condition in many parts of the world. Droughts in some places and unseasonably heavy rain in some other places are attributed to El Nino.

La Nina:
La Nina is some times called anti-El Nino. It represents periods of below average sea surface temperatures along the east-central equatorial Pacific. Its global climate impact tends to be opposite to that of El Nino. In some parts of the world, it causes increased rainfall, while in other regions it causes extreme dry conditions.

Montreal Protocol:
Finalized in 1987, it is a global agreement to protect the stratospheric ozone layer by phasing out production of ozone depleting substances such as Hydro chlorofluoro carbons (HCFCB's) and Chloro fluorocarbons(CFCB's).

Paris Climate Agreement:
It is an accord within the United Nations frame work dealing with green house gas emission mitigation. The agreement will come into effect starting from the year 2020 and is negotiated between 196 participating countries, whose representatives attended the conference in Paris in December 2015. The Paris agreement adopted by consensus aims at limiting the global temperature rise to less than 2°C from the pre-industrial level and to pursue efforts to limit it even further to 1.5°C.

Intended Nationally Determined Contributions (INDCS):
It is a term used at the December 2015 Paris conference specifying individual countries' strategy in reducing GHG emissions. Under the Paris agreement China and USA ratified the 2015 agreement on INDCS in August 2016. Together, they both contribute 38% of the total global emissions, with China alone emitting 20%. India which has a global share of 4.1% of Green House Gas (GHG) emission ratified the Paris climate agreement in October 2016. In June 2017, President Donald Trump announced his intention to withdraw USA from the agreement. The earliest effective date for withdrawal for the US, is November 2020.

Neuron Cell:
The human body has about 200 different types of cells. One such group of cells are responsible for transmitting sensations such as pain, pleasure etc, are known as neurons. A typical neuron cell has a cell body called the nucleus with two or more long fibers called *dendrites* which carry signals into the cell body. Only one cell fiber called the *axon* carry the signals away from the cell body. Bundles of fibers from the neurons are held together by connective tissues which in turn form nerves. The human brain is estimated to have about 8.5 billion neurons. Neurons while connecting together do not touch each other, and instead forms tiny gaps called *synapses* at their end. These gaps can be either chemical or electrical synapses and pass signals from one neuron to the next.

Artificial Neural Networks:
In artificial neural networks, an attempt is made to simulate the brain function. An artificial neuron as opposed to the biological neuron consists of a simple electronic device which receives signals from various sources, and are then weighted as per requirements and finally added together. If the weighted sum of the signals exceeds a certain value, called the threshold value, then the neuron fires. Otherwise, no firing takes place. These individual neurons are connected together to form an artificial neural network.

Moore's Law in Electronics:
This law refers to an observation made by Intel co-founder Gordon Moore in 1965. He noticed that the number of transistors per square centimeter on integrated circuits gets doubled every 18 months. Moore's law predicts that the trend will continue into the foreseeable future, although the pace has slowed down in recent years. As an extension of Moore's law, we can surmise that computer response will become faster in the future till inter atomic distances are reached.

Murphy's Law:
It is an epigram that typically states that "Anything that can go wrong will go wrong" and one may further add "usually at the worst time".

Occam's Razor:
This principle states that when presented with competing theories to explain a physical phenomenon, the simplest explanation is more likely to be correct. The idea is attributed to William Occam, a fourteenth century Anglo-French theologian. It serves as a heuristic guide in the development of theoretical models.

Rational Numbers:
They are numbers that can be expressed by dividing one integer by another non-zero integer. When expressed in decimal form, rational numbers either come to an end after a finite number of digits (e.g 8/5 = 1.6) or when certain digits are repeated for ever (e.g 1/3 = 0.333……).

Irrational Numbers:
They represent numbers that cannot be expressed by dividing one natural number by another. Unlike rational numbers they cannot be expressed as a ratio between two integers. Some of the most important numbers in mathematics are irrational, e.g. the ratio between the circumference of a circle and its diameter, denoted by π.

Other examples are Euler's constant e, and square root of 2 denoted by $\sqrt{2}$.

Algebraic and Transcendental Numbers:
An algebraic number is one that can be obtained as a solution of a polynomial equation in a single variable x with rational coefficients. A transcendental number cannot be obtained in this manner. For example, a polynomial $x^2 - 2 = 0$, which has rational coefficients 1 and 2, has a solution $x = \sqrt{2}$, which is an irrational number but not a transcendental number.

Examples of common transcendental numbers are $\pi = 3.14159...$ and $e = 2.71828...$

Prime Number:
Is a positive integer greater than 1, whose only divisors are 1 and the number itself. For example, 2, 5, 7, 11, 13 …….etc.

Gödel's Incompleteness Theorems:
These theorems of mathematical logic demonstrate the inherent limitations of every formal axiomatic system. Published by Kurt Gödel in 1931, they are important in mathematical logic. Two theorems are stated in this connection. The first incompleteness theorem states that in any consistent formal system F, within which a certain amount of arithmetic can be carried out, there are statements of the language F which can be neither proved nor disproved in F. The second incompleteness theorem states that such a formal system cannot prove that the system itself is consistent.

Index

A

algebraic and transcendental numbers, 168
alpha decay, 40
amino acids, 98
ampere's law, 39
Anderson, Paul, 24
anti-particle, 36
Aristotle, 9, 20
artificial intelligence, 4, 121
artificial neural network, 167
asteroid, 54
astronomical unit, 152
asymptotic freedom, 40
atomic mass number, 22
atomic mass unit, 22
atomic number, 22
axon, 123

B

baryon, 36
base pairing principle, 105
beta decay, 42
binary pulsar, 33
Binding energy, 69, 70
Big Bang, 10
black body radiation, 87, 161
Black hole, 32
bose – einstein statistics, 26, 37
bohr model of the atom, 89
Bohr, Niles, 20
Born, Max, 87
brownian motion, 79

C

cephid variables, 154
Ceres, 47
Cern, 26
chandra x-ray observatory, 61
chandrasekhar limit, 31, 154
Chandrasekhar, Subramaniam, 31
chernobyl, 77
chloro fluro carbon, 165
Chomsky, Noam, 122
chromosomes, 100, 107
codon, 98
comet, 53
comton gamma ray observatory, 61
Copernicus, Nicholas, 9
cosmic background explorer, 11, 58
cosmic microwave background radiation, 11
Crick and Watson, 98
CRISP – Cas 9, 113, 114, 165
critical density, 13
cytoplasm, 107

D

Dalton, John, 20
dark energy, 12
dark matter, 12
Darwin, Charles, 97
Dawkins, Richard, 148
DeBrogli, Louis, 20, 89
Democritus, 20
dendrite, 123

Deoxyribo nuclic acid, 162
Deuterium, 36
Dirac, Paul, 23, 87
doppler effect, 152
double helix structure, 98
dwarf planets, 52

E

ecliptic plane, 45
Edington, Arthur 4, 148
Einstein, Albert, 37, 38, 150
electro - weak force, 42
electron degeneracy, 31
electron volt, 21
elementary particles, 26
El Nino, La Nina, 166
embryonic stem cells, 108
energy levels, 21
epigenetics, 103, 164
Eris, 47
event horizon, 155
exo planet, 54
expert system, 126

F

Faraday's law, 39
Fermat's last theorem, 144
Fermi – dirac stastistics, 26, 37
fermion, 36
feynman path integral, 94
Feynman, Richard, 94
Fukushima, 376
fuzzy logic, 128

G

galaxy, 152
Galileo, 4
gamma ray burst, 155
Gellman, Murray, 24

gene, 98, 162
genetic algorithm, 127, 128
genome, 106
Glashow, Sheldon, 26
global warming, 115
gluon, 26, 35
God particle, 28
Godel's incompleteness theorem, 168
Graham, Bell, Alexander, 132
grand unified theory, 27
gravitational lensing, 84, 160
gravitational waves, 63
graviton, 35
greenhouse warming, 117, 165

H

Hadron, 36
Haldane, JBS, 97
Hawking, Stephen, 128
Heisenberg principle of uncertainty, 31, 161
Heisenberg, 3
Higgs particle, 26, 62
Hubble, Telescope, 55, 59
Hubble, Edwin, 9
hubble's constant, 152
Hubble's law, 14
human genome project, 3

I

inertial frame of reference, 80
insulation, 153
intelligent design, 148
international astronomers union, 46
international experimental thermonuclear reactor, 75
international space station, 61
inverse beta decay, 32
ionisation energy, 22

isolated pulsar, 33
isotopes, 22, 68

J
James Webb telescope, 55, 60
James, Chadwick, 31
Jeans, James, 31

K
Keck telescope, 58
Kepler, Johannes, 9
Kepler's laws, 54
Kepler's space astronomical lab, 61
Kuiper belt, 45, 154

L
Lagrange point, 60
large hadron collider, 27, 62
Laser Interferometer Gravitational observatory, 63, 157
Lemaitre, George, 15
length contraction, 18, 81, 82
lepton, 36
light water reactor, 73
light year, 152
Lorenz contraction, 160

M
mass defect, 69
mass energy equivalence, 83
Maxwell's equation, 38
McCarthy, John, 121
Medawar, Peter, 111, 149
muon, 23
Mersenne number, 143
messenger RNA, 102
meteroid, metor, meteorite, 155
mitocondria, 108
Moore, Gordan, 132
Moore's law, 131, 167

morse code, 132
Murphy's law, 167

N
nebula, 154
neuro transmitter, 123
neuron cells, 167
neutrino astronomy, 63
neutrino, 23
neutron observatory, 157
neutron star, 32,154
nucleon, 35
nucleosynthesis, 159

O
Occam, Razer, 167
Oort, Cloud, 146, 154
optical fiber communication, 133
orbiter mission, 64, 65
ozone layer, 165

P
parsec, 152
Pauli, Wolfgang, 23, 93
Pauli's explosion principle, 31, 36
photoelectric effect, 88
photon, 35
Planck constant, 88
planck length, planck time, 160
Planck, Max, 87
pluripotent cell, 110
pluripotent cell,100, 109, 164, induced
Popper, Karl, 147
positron emission, 42
positron, 23
pressurized heavy water reactor, 73
prime number theorem, 142
Ptolemy, 9
pulsar, 32, 155

Q

quantum chromo dynamics, 26, 40
quantum electro dynamics, 26, 40
quantum number, 90
Quark, 24
Quasar, 34, 155

R

Rabi, Isdor Isaac, 23
red giant, 31
red shift, 152
Rees, Martin, 15
retro virus, 112, 164
Ribo Nuclic Acid, 163
ribosome, 102, 163
robotics, 127
Rosalind, Franklin, 98
Russel, Bertand, 6, 139
Rutherford, Earnest, 20

S

Salam, Abdus, 26
Schrodinger equation, 93, 161
Schrodinger, Erwin, 3, 88
Schwarzschild radius, 34, 161
science literacy, 5
scientific temper, 5
SegattArius A, 34
Shockly, Bardeen and Brattain, 134
Snow, CP, 1
Somatic Cell Nuclear Transfer, 110
standard candle, 32
standard module of particle physics, 24
stem cells, 164
strong force, 40, 68

Supernova, 32, 152
synaptic sac, 123

T

Tau particle, 24
Teleomeres, 101, 102, 163
Thomson, JJ, 19
Three Mile Island Disaster, 76
tidel locking, 48
time dilation, 80, 81, 161
Tokamac, 74
transistor, 134
tritium, 36
troposphere, 116
Turing test, 122
Tycho, Brahe, 9

U

Universal Law of Gravitation, 38

W

W and Z particles, 26, 35
Watson, J, 3
Wave particle duality, 88
weak force, 40
Wheeler, JA, 38
white dwarf, 31, 153
Wien's displacement law, 162
Wienberg, Steven, 26
work function, 88
x-ray pulsar, 33

Z

Zadeh, Lofti, 128
Zygote, 99

www.ingramcontent.com/pod-product-compliance
Lightning Source LLC
Chambersburg PA
CBHW030943180526
45163CB00002B/685